Citizens, Experts, and the Environment

FRANK FISCHER

Citizens,

Experts,

and the

Environment

THE POLITICS OF LOCAL KNOWLEDGE

DUKE UNIVERSITY PRESS DURHAM AND LONDON 2000

© 2000 Duke University Press

All rights reserved   Printed in the United States of America on acid-free paper

∞   Designed by C. H. Westmoreland   Typeset in Sabon with Twentieth Century display by Keystone Typesetting, Inc.   Library of Congress Cataloging-in-Publication Data appear on the last printed page of this book.

For Elizabeth Fischer

Contents

Preface

The tension between professional expertise and democratic governance is an important political dimension of our time. Democracy's emphasis on equality of citizenship, public opinion, and freedom of choice exists in an uneasy relationship with the scientific expert's rational, calculating spirit. At times, especially in cases pertaining to science, technology, and the environment, the tension breaks out into open conflict. Think, for example, of the contemporary challenges of the religious creationists to the scientific teaching of biological evolution, or the growing public rejection of genetically engineered foods. Some writers even suggest that the division between those with and those without expert knowledge will be one of the basic sources of social and political conflict in the new century.

Often such conflicts result from the overapplication of scientific rationality to public policy making. Concerns about the apolitical character of technocratic modes of thought and action have emerged as critical social questions in the second half of the twentieth century. Fundamental to these concerns is the role of democratic participation in an increasingly expert-driven society: do most citizens have the knowledge and the intellectual wherewithal to contribute meaningfully to the complex policy decisions facing an advanced industrial society? The question poses a challenging issue for democratic theory and practice.

Everyone, at least officially, is for democracy. The call for citizen participation is a prominent theme in both public and academic discussion. The United States, moreover, spends enormous sums of money each year to promote democracy around the world. Despite the enthusiasm for promoting democracy, most of the calls for more democracy and citizen participation occur at the same time that we witness the disturbing decline of democratic practices at home. Not only do the attitudes and behaviors of citizens reflect a growing apathy toward their political institutions, but the political leaders of the Western democracies all too often seek to circumvent the democratic process.

Many factors explain the wane of the citizen's role. Among the most important is the social and technical complexity of modern societies. What are the possibilities, many ask, of the ordinary (that is, nonexpert) citizen deliberating intelligently on the policy issues confronting the decision makers of such societies? Are not these issues better addressed by the professional experts? Given that citizen participation is one of the foundations of a strong democracy, such arguments may give us pause.

Democracy, of course, means different things to different people. Participatory democrats call for more citizen participation, understood as deliberation on the issues affecting their own lives. For such democrats, participation not only gives meaning to democracy but also plays an important educational and psychological role in the social development of the individual citizen. Others, skeptical of the rational capabilities of the ordinary citizen, lend their support to representative government, preferring to limit the citizen's role to elections and voting. Still others, usually less openly, argue that genuine citizen participation is a matter whose time has passed. Skeptics, moreover, make much of the fact that most citizens seldom participate in the political process. Although the reasons for this lack of interest are less than obvious, the point can scarcely be ignored.

Some portray the situation as a dilemma: citizens don't have enough knowledge to participate meaningfully in technically oriented policy decisions, but it is difficult in a democracy to legitimately deny citizens a place at the decision-making table. Indeed, it is inevitable that people will continue to demand to have a say. This has led many scientists and politicians to see citizen participation caught in a dilemma between impossibility versus inevitability (Doble and Richardson 1992).

But is this in fact a hopeless dilemma? Is it possible to innovate new forums that can constructively circumvent what may otherwise be a standoff between citizens and experts? This is the question addressed in the pages that follow.

Part of the difficulty in answering this question rests with political and social inquiry. Despite the contemporary emphasis on citizenship, democratic theorists largely remain distant from the level of the citizen. As Blaug (1998) points out, such theorists mainly labor at the abstract level of the nation-state and, in doing so, neglect the everyday aspects of deliberative politics, especially as they relate to ordinary people. Even the contemporary emphases on the public sphere and communitarian practices largely focus on the contributions of citizen deliberation to the legitimacy and meaning of political institutions rather than to citizen groups.

In an effort to bring the question down to the level of the citizen, this book inquires into the *realistic* possibilities of meaningful citizen participation. Moving beyond the standard ideological exhortations for or against citizen participation, the study does this by confronting one of the hardest of issues, social and technical complexity. Focusing on the complex questions posed by environmental risk, the question is this: what evidence supports the contention that citizens can effectively participate in helping to make the complex decisions facing contemporary policy makers?

The question is approached both theoretically and pragmatically. Although participation is a political virtue in and of itself, I acknowledge from the outset that it can be a frustrating endeavor. Collective citizen participation is seldom something that simply happens. To succeed, it often has to be organized, facilitated, and even nurtured. Experience shows, moreover, that participation is highly useful in some cases and not in others. For this reason, I seek to problematize the question of participation. Drawing on a wide range of cases, this work carefully assesses what citizens *can* in fact do, what kinds of institutional reforms will help them do that, and in which kinds of policy domains such participation is useful. Toward this end, the discussion draws heavily on lessons from the practice of participatory inquiry.

The picture presented here challenges the conventional wisdom. Hard evidence demonstrates that the ordinary citizen is capable of a great deal more participation than generally recognized or acknowledged. At the same time, while the analysis supports the case for participatory democracy, it does not present citizen participation as a magic cure-all for economic and social problems. Nor is deliberation or argumentation meant to direct attention away from questions of interest and power. But it does hold out the possibility of bringing forth new knowledge and ideas capable of creating and legitimating new interests, reshaping our understanding of existing interests, and, in the process, influencing the political pathways along which power and interests travel.

Specifically, the book examines the ways in which the deliberations of ordinary citizens can have an important—even, at times, essential—impact on environmental problem solving. Not only can they help in searching for solutions to pressing environmental problems, but they can also contribute a kind of knowledge—in particular, local knowledge—that the professional expert requires. Furthermore, the analysis illustrates how the case for local knowledge is buttressed by insights from contemporary epistemology and the sociology of science. The work sets out ways by which the relationship between citizens and experts can be restructured to better facilitate an epistemological integration of both expert and citizen knowledge(s). Finally, it looks at the kinds of institutional reforms that can encourage and facilitate such practices.

The analysis is developed in four parts. Part 1 sets the stage with a general discussion of the role of technology and expertise in modern society. Taking up the case for citizen participation, this first part places the tension between citizens and experts in the context of environmental risk. The first three chapters analyze participation as both an ideology and a practice, examine the critique of professional expertise, and explore the call for alternative practices. The fourth chapter presents the epistemological issues underlying the critique of scientific expertise and offers a postpositivist, discursive theory of knowledge. Challenging the scientific expert's methodological emphasis on "generalizable knowledge," postpositivist theory underscores the importance of bringing in the local contextual knowledge of the ordinary citizen. In this sense, the case for participation is seen to be as much grounded in epistemology as in democratic politics.

Turning to concrete issues, part 2 focuses on the role of scientific expertise in environmental policy making, the political response of the environmental movement to technocratic decision practices, and the resulting politics of counterexpertise. Specifically, it considers the conflict between the technical rationality of the environmental expert and the sociocultural orientation of the citizen environmentalist. The challenge is seen to be how to discursively integrate the two forms of knowledge, rather than rejecting one for the other.

Part 3 explores the deliberative alternative, examining both the practices of participatory inquiry and particular case experiences with such research practices. It also explores the concept of local knowledge, the primary product of participatory inquiry, and demonstrates its implications for both environmental risk and policy analysis generally. Part 4 then concludes with an illustration of how the participatory inquiry of lay citizens and experts can be brought to bear on complex decisions about environmental risk and the ways that their collaborative evaluations can be used to inform national legislators. These concluding chapters assess the prospects of a more discursive, participatory mode of policy expertise and outline the need for the study of "policy epistemics," an approach to knowledge designed to help better understand the relationship between citizens and experts. To this end, the work calls for a new understanding of the expert as "specialized citizen."

Finally, it is important to thank a number of people, who, during the writing of this book, were kind enough to provide me with useful comments on various parts of the manuscript. These include Douglas Torgerson, Susan Fainstein, Vatche Gabrielien, Alan Mandell, Maarten Hajer, Dvora Yanow, Robert Hoppe, Wayne Parsons, Patsy Healey, Herbert Gottweis, Jeanette Hofmann, Simon Joss, Hubertus Buchstein, Judith Innes, Marion Nestle, Michel van Eeten, Daniel Barben, Mona Choudhary, Navdeep Mathur, Patria Delancer, Martin Benninghof, Laura Solitare, Rina Majumdar, Mark Brown, and N. C. Narayanan. Although none of them bear any responsibility for what follows, all provided me with useful advice at various points along the way. Last but certainly not least, I am especially grateful to Valerie Millholland of Duke University Press for providing steady editorial support and encouragement. To all of the foregoing, I extend my heartfelt thanks.

Two of the chapters in this book are elaborations of earlier essays. Chapter 4 is a revised version of "Beyond Empiricism: Policy Inquiry in Postpositivist Perspective," in *Policy Studies Journal* 26, no. 1:

129–46. An earlier version of chapter 9 appeared in my book *Technocracy and the Politics of Expertise* as "Restructuring Practice: The Elements of Participatory Practice," Sage Publications, 1990. In addition, Appendix A is drawn from my 1995 book, *Evaluating Public Policy,* published by Wadsworth Publishing Co. Appendix C is taken from "Farmers as Analysts, Facilitators, and Decision-Makers," by Parmesh Shah in *Power and Participatory Development: Theory and Practice,* ed. Nici Nelson and Susan Wright, Intermediate Technology Publications, 1995. Appendix D comes from "Participatory Research as Critical Theory: The North Bonneville, USA Experience," by Donald E. Comstock and Russell Fox in *Voices of Change: Participatory Research in the United States and Canada,* ed. Peter Park, et al., Toronto: OISE Press, 1993, 101–24. And Appendix E is excerpted from Dan Durning, "Participatory Policy Analysis in a Social Services Agency: A Case Study," *Journal of Policy Analysis and Management* 12, no. 2, 297–322. I wish to thank the publishers of these works for having granted me permission to republish these materials.

PART I

Citizens and Experts in the Risk Society

. . . whenever the people are well-informed, they can be trusted
with their own government . . . — Thomas Jefferson

Can democracy thrive in a complex technological society? The question is one of the most challenging political issues of our time. Although democracy remains an undisputed public ideal, all the more so
after the Cold War, the prospects of meaningful citizen participation
in an age dominated by complexity and expertise are neither clear nor
obvious. Rhetorical appeals to democracy aside, citizen participation
in Western democracies has fallen to disturbingly low levels, especially
in the United States. Given this present or even worsening trend, the
very term threatens to lose its meaning. Already talk of democracy all
too often serves as little more than a thinly veiled guise for elite governance. The question is, Can the democratic process be rescued from
the increasingly technocratic, elitist policy-making processes that
more and more define our present age?

Citizen participation, defined as deliberation on issues affecting
one's own life, is the normative core of democracy. Even though
widely accepted as a basic political value, citizen participation often
remains controversial. Mostly everyone is for it, at least in principle,
but many are quite skeptical of it when it comes to specific issues or
practices. Given the state of public opinion and citizen participation,
many argue that it is better to rely on the experts. Are not the knowledge elites more likely to support the values of both social justice and
efficiency, they ask? Given the racist, nationalistic, and class-based

attitudes that pervade Western political systems, are we not wiser to invest our trust in the experts?

Why, then, citizen participation? Broadly speaking, participation contributes to three important goals. First of all, citizen participation and its normative rationale, deliberation, give meaning to democracy. If we are to take seriously a "strong" form of democracy, as Barber (1984) puts it, all citizens need to deliberate at least some of the time on the decisions that affect their lives. Second, citizen participation contributes normatively to the legitimization of policy development and implementation. And third, but not least important, citizen participation can contribute to professional inquiry. Participatory forms of inquiry, as will be seen here, have the potential to provide new knowledge — in particular local knowledge — that is inaccessible to more abstract empirical methods.

None of this is to suggest that citizen participation is the magic solution to our pressing social and economic problems. Rather, it is to argue that citizens can and should contribute to the search for the solutions to these problems. Toward this end, the overarching purpose of this work is to outline new approaches for bringing citizens and experts together. Before doing that, however, it is necessary to set the stage with a more general discussion of the role of technology and expertise in modern society, the possibilities of citizens' participation, and the presentation of the particular political issue to be used to make a theoretical issue concrete. No other issue better illustrates the nature of this problem than environmental risk. As a way of making concrete and practical an otherwise theoretical question, the chapters that follow examine the issue of expert knowledge and citizen participation in the political context of environmental risk — or what Wildavsky (1988) called "the pursuit of safety."

The goal of the first four chapters is to set out a foundation for the more systematic treatment of the issues in the rest of the book. Chapter 1 focuses on the central role of technology and scientific expertise in modern society and their implications for democratic decision making. Emphasizing polemics that have taken place mainly in the United States, the discussion focuses critically on the technocratic form of expertise and its instrumentalization of reason. In particular, the importance of technocratic expertise is lodged less in the central position of the expert in the decision-making structures than in the impact of expert discourses on the way we understand and organize the world. Toward this end, Foucault's analysis of professional "disciplinary powers" is introduced. Acknowledging the normative problems of agency and social reconstruction posed by his poststructural perspective, his concept of local resistance is emphasized. The chapter concludes by setting out the retrieval of local knowledge and its relation to the politics of reconstructing professional discourses as the central concerns of the task that follows.

Chapter 2 turns to the question of citizen participation. Here the fundamental public challenges facing contemporary professionals are raised — charges from the lay citizenry that experts are more concerned with their own wealth and status than in the interests or safety of the public. Within the disciplines, this has led some to begin rethinking the nature and conduct of professional practices. One of the most significant alternatives has been that of advocacy research and its politics of counterexpertise, but this practice falls short in terms of citizen participation per se, leading others to call for more participatory forms of inquiry.

The second part of the chapter asks if citizens are actually able to take a more participatory role in the complex decisions of our time. How do we explain the fact that participation is low? Does this mean people are uninterested in, or incapable of, participating? Or does it reflect the fact that existing political structures present most of us with few meaningful opportunities? The chapter concludes that while we don't really have solid answers to these questions, there is a good deal of evidence to suggest that citizens are capable of participating much more than the conventional wisdom would have us believe. The challenge for political science and sociology is to explore the boundaries of the possible.

Chapter 3 introduces environmental politics as a context for illustrating more concretely the nature of the conflicts between citizens and experts and the question of participation more generally. For this purpose, the chapter examines the techno-industrial logic of Beck's "risk society," along with its accompanying politics of expertise and counterexpertise. Beck focuses on the fact that the public more and more recognizes the environmental dangers accompanying industrial progress to be the result of corporate and state institutional decisions. This ushers in, he argues, a new political questioning of the nature of

3

modern technological society itself. Basic to this political reflexivity are emerging tensions between scientific experts and the lay public.

As the growing influence of science and technology gives rise to increasingly public fears and disputes about its privileged status, laypersons express political uncertainty and hesitation about the implementation of scientific and technological projects — from nuclear energy to biotechnology. More and more environmental groups, citizens, and politicians speak of the need to regulate and control science. While the scientific community complains of intervention in the pursuit of knowledge, the public increasingly comes to see that scientists are themselves laypersons in matters concerning political goals and social judgment. Bringing these scientific and normative judgments together requires new institutional forums. For Beck, the solution is to be found in a more participatory form of democracy, or what he calls "ecological democracy." But Beck tells us little about how this ecological democracy might bring together citizens and experts. It is to the issues and problems involved in bringing about such a participatory democracy that much of this book is devoted.

As the latter half of chapter 3 argues, the democratic restructuring of scientific decision making and the interactions between citizens and experts involves more than new institutional forums. Even more fundamentally, such a restructuring necessitates a critical reexamination of the concept of knowledge itself. Chapter 4 turns to this question. First, the discussion offers a critique of the dominant neopositivist conception of science underlying both the conventional practices of scientific inquiry and professional expertise. It then presents a postpositivist framework for practical deliberation that is designed to bring together citizens and experts in a more participatory mode of mutual inquiry. Based on a constructivist conception of social reality, the approach turns from the traditional emphasis on scientific proof to a contextually oriented discursive understanding of social inquiry. In doing so, it discursively situates knowledge in the context of time and local circumstances. The discussion lays the groundwork for the more extensive discussion of local knowledge and forms of citizen-oriented participatory inquiry in parts 3 and 4.

1. Democratic Prospects in an Age of Expertise

Confronting the Technocratic Challenge

Much of the history of . . . progress in the Twentieth Century can be described in terms of the transfer of wider and wider areas of public policy from politics to expertise. — Harvey Brooks

The language and iconography of democracy dominates all the politics of our time, but political power is no less elitist for all that. So too the technocracy continues to respect the formal surface of democratic politics; it is another, and this time extraordinarily potent means of subverting democracy from within its own ideals and institutions. It is a citadel of expertise dominating the high ground of urban-industrial society. . . . — Theodore Roszak

Everywhere in the world, democratic institutions are gaining new adherents, with American democracy widely seen as the model to emulate. In the flush of such post Cold War enthusiasm, however, the fact that U.S.-style democracy has been experiencing its own troubles has too often been overlooked. To be sure, there have been no short-ages of analyses of such problems: Why do so many Americans show such little interest in voting? Why do they hold their political institu-tions in such low esteem (Dionne 1991; Barber 1984)? Why has the level of public discourse devolved to that of simplistic television com-mercials (Bennett 1992)? And so on. Following in this line of inquiry, this work seeks to take up a critical aspect of the question that has received far too little systematic attention; namely, how can citizens participate in an age dominated by complex technologies and expert decisions (Fischer 1990)? Indeed, no other aspect of the contemporary "democracy question" can be more important.

The basic question I pose here is scarcely new. In the 1920s, John Dewey (1927) forcefully raised it in his book *The Public and Its Problems.* Engaging the challenge to democratic governance in the emerging twentieth-century industrial society, Dewey asked how a mass public could deal with the increasingly complex nature of the problems presented by a highly differentiated, technologically driven society. How could citizens participate in political decision making so obviously dependent on the knowledge of experts?

Indeed, Dewey identified a paradox. As the importance of the citizen grew in the political realm — thanks to the expansion of basic rights in the nineteenth and twentieth centuries — the phenomenon was paralleled by the growth in power of large corporate and governmental organizations directed by managerial and technical expertise. Thus in just that period in which the political influence of the citizenry was taking shape, it was undercut by the rise of bureaucratic organization and technical expertise.

Large-scale industrial society transformed the very nature of everyday life. No longer did most people provide their own necessities — grow their own food, supply their own means of transportation, build their own dwellings, and so on. In industrial society all these basic goods and services are mass-produced and marketed through large, highly interdependent, impersonal structures and functions ever-increasingly dependent on expert systems.[1] Given these features of industrial society, in particular the central role of expertise, Dewey saw little future prospect of well-integrated political communities organized around a knowledgeable citizenry. Under these new social arrangements, individual citizens could no longer easily comprehend the processes through which their daily needs were satisfied. As a consequence, they could no longer be expected to easily determine their own interests. Such a situation, he worried, could lead large numbers of citizens to embrace simplistic or false ideas. In their search for social reassurance, such citizens could easily fall victim to ideas antithetical to democracy, fascism and communism being the primary twentieth-century examples.

6

How is it possible to overcome the challenge posed by this un-precedented level of social and technical complexity? The answer for Dewey was a division of labor between citizens and experts. On the technical front, experts would analytically identify basic social needs and problems. On the political front, citizens could set a democratic agenda for pursuing these needs and troubles. To integrate the two processes, Dewey called for an improvement of the methods and conditions of debate, discussion, and persuasion. Indeed, in his view, the need for such improvements was *the* problem of the public. Debate would require the participation of experts, but they would act in a special way: instead of rendering judgments they would analyze and interpret. If experts, acting as teachers and interpreters, could decipher the technological world for citizens and enable them to make sensible political judgments, the constitutional processes designed to advance public over selfish interests could function as originally intended.

Since Dewey's time, the progress of democracy has been disappointing. Although Western democracies exhibit high degrees of interest group involvement, levels of individual citizen participation (as ample political and sociological research shows) have declined rather than expanded. While the interpretation of this phenomenon is complicated, as we will see in chapter 2, it has led many to question the very capacity of the citizenry to render judgments on the complex issues that define our times.

Over the same period, moreover, professional experts have failed to ease the problem. Rather than adopting the role of teacher or educator, as Dewey had hoped they would, experts have largely set themselves off from the mass citizenry. Instead of facilitating democracy, they have mainly given shape to a more technocratic form of decision making, far more elitist than democratic. Dahl (1989, 337) captured this concern in his assessment of existing democratic arrangements. The increasing complexity of social problems, giving rise to increasing specialization and the expansion of elite "public policy specialists," puts the Western polyarchies in the position of being replaced by a "quasi-guardianship" of autonomous experts, no longer accountable to the ordinary public.

To make matters worse, over the past decades we have come more and more to learn that the experts are themselves incapable of answering these questions. Not only do experts lack answers to the complex technical questions that confront us, but expertise itself turns out not to be the neutral, objective phenomenon that it has purported to be. Indeed, it has all too often served the ideological function of legitimating decisions made elsewhere by political rather than scientific means. Nowhere, as we shall see, has this phenomenon been more evident than in environmental politics, one of the most technologically driven of the policy fields.

In recent years, this concern with both the complexity and uncertainty of our problems has lead influential political theorists to rethink their positions on the prospects of democracy. For instance, Bobbio (1987) sees the project of political democracy as being unable to fulfill its promises in complex modern societies. The expanded involvement of technical experts in policy making and the increasing process of bureaucratization serve as major structural obstacles to the fulfillment of the original democratic ideal.

Similarly, complexity is one of the main issues that troubles Habermas in his ongoing effort to spell out a theory of deliberative democracy. Whereas in his early writing Habermas shared Dewey's optimism about the possibilities of meaningful citizen participation, he has adopted a more pessimistic tone. In his view, "unavoidable complexity" imposes the need for important qualifications in the elaboration of participatory democracy. Shifting away from his earlier theory of radical democracy, Habermas suggests that democracy may not apply to all realms of decision making. This, of course, remains a contested theory. But the fact that one of the leading political theorists of our time has decided that the evidence suggests that the complexity of modern societies poses constraints on the full democratization of societal decision making calls to attention the seriousness of the question.

The central goal of this work is to put the relationship between citizens and experts together in a new way, one capable of making good on Dewey's initial proposal. As the effort depends on an understanding of both political and epistemological developments that have evolved in more recent times, the foundations of the position need to be developed before addressing the solution directly. Toward this end, the rest of this chapter is devoted to complexity and the rise of expertise, focusing in particular on technocratic politics and its implications for democracy and citizen participation.

Complexity and the Age of Expertise

To be sure, one of the classical questions of political science and sociology has been the issue of the relation of knowledge to power (Fischer 1990, 59–76; Ezrahi 1990). Much of the writing about this problem, however, has been theoretical. When translated into the more practical question of the relation of the citizen to the expert, our knowledge of the relationship remains disturbingly inadequate. This is especially unfortunate, given that in our highly complex technological society, the experts have moved to center stage.

In this age of expertise, the question of knowledge and competence cuts across the entire spectrum of political and governmental issues. For this reason, policy questions today present the complicated task of not only coming to grips with expert analyses of sophisticated technical issues but also understanding how different citizens arrive at their own judgments about such issues, including their understandings of the experts themselves. Moreover, the increasing unwillingness of citizens to accept uncritically the trained judgments of the experts has become one of the central issues of our time. Indeed, as we shall see, it is one of the primary political dynamics of environmental decision making (Hays 1987, 329–62).

Such conflict between citizens and experts poses a dilemma. The need for specialized expertise bears directly on how much citizens can know about the choices they confront. Not only does this directly involve the technical dimensions of policy questions, but it concerns as well the value trade-offs and other consequences that follow from the implementation of such policies (Hill 1992). Decision-making procedures, in this respect, must take into consideration the authority and influence that different actors have on the final choices. Should such decisions be left to the experts? What level of influence, for example, should the views of the general public carry when compared, for example, to those of scientists, administrators, elected officials, engaged community leaders, and activists? Who is more capable of judging whether a power plant or a new regulatory program serves the interests of the public?

How we devise solutions to these questions is structured by our assumptions about citizens' cognitive abilities to participate in discussions about complex issues, including their methods of assessment. Such assumptions, unfortunately, are often based on fundamental ideological perspectives — political liberals typically call for more public involvement; conservatives, for less. As such, the issue is bound up as much with competing interests and ideologies as with well-founded evidence.

The next two sections outline these two rival perspectives, that of progressively oriented liberals and that of the political conservatives. One of the central issues in the contemporary variant of this debate is the role or significance of an overarching concept, the "information society" (Lyon 1988). It offers a useful wedge into this discussion.

Expertise and the Information Society

The Celebration of Technology

The most important contemporary expression of the central role of expertise has been the discussion of the "postindustrial society" and its latest variant, the "information society" (T. Forester 1985; Poster 1990, 21–42; Luke 1990). Both terms designate a social formation in which the codification and use of knowledge become fundamental organizational principles of society (Bendiger 1986). The reproduction of "information value" rather than "material value" is seen increasingly to be the driving force of this new societal formation. Information value gives rise to industries based on the computer sciences, telecommunications, robotics, and biotechnology (concerned with breaking the information code of life itself). These burgeoning "information industries" are widely recognized as transforming the very economic and social fabric of Western societies (Castells 1996). Today their symbol has become the computer and the so-called information highway connecting computers throughout the world.

The arrival of the information society is much celebrated in many elite circles, economic and political as well as intellectual. Distinguished management writer and guru Peter Drucker sees the dramatic expansion of information as ushering in a profoundly important new era with unprecedented societal implications (Drucker 1993). Arguing that the information or knowledge society has already created a postcapitalist society promising a global transformation, he writes that "knowledge is the only meaningful resource today" (Drucker 1993, 65). The "traditional factors of production—land, labor, and capital—have not disappeared, but they have become secondary. They can be obtained, and obtained easily, provided there is knowledge" (also page 65). For Drucker, information is an objective utility for achieving desired economic and social outcomes. Its application, in his view, is the essence of innovation.

As Drucker focuses on the economic aspects of this new information era, so the futurist Alvin Toffler regularly writes and lectures on the social and cultural implications. He describes this "Third Wave" as a system radically changing the nature and extent of human interactions. The offspring of the union between computing and telecommunications, the information society promises an all new kind of society. As Toffler puts it, "what is now occurring . . . is in all likelihood bigger, deeper and more important than the industrial revolution. . . . the present moment represents nothing less than the second great divide in human history" (Toffler 1993, 21).

In his books, Toffler (1991; 1993) outlines a vision of national transformation through information access. In his view, we are entering the "Knowledge Age," as part of the "Third Wave" of history. The new wave ushers in a society in which knowledge replaces matter — natural resources or energy — as a source of power. It is a power that will flow through interconnected computers, or in "cyberspace." The central role of societal institutions in this new formation is to remove the barriers that hinder or impede the shift of information from institutions to individual citizens. Government is advised to remove "second wave" laws and regulations from the telecommunications and computer industries. The absence of restrictions will return power to self-reliant citizens; we are seen to stand on the threshold of a brave new libertarian future.

While some of this is surely armchair philosophy, its political importance should not be underestimated. Both of these writers were at the top of the reading list Newt Gingrich offered to his fellow Republicans as they took over the U.S. House of Representatives in 1994 and initiated their efforts to advance a new social "Contract with America." Gingrich described these works as the source for inspiration and renewal in twenty-first-century America. Early into his term as

Speaker of the House, Gingrich unveiled an electronic archive on the Internet that contains every bill submitted to Congress and every speech uttered in the House and the Senate. According to Gingrich, the archive is just one part of an ambitious plan to take power away from the elites and give it back to the people. As Bennahum (1995, A23) explains it, "Gingrich's goal is a return to a Jeffersonian past with a 21st-century twist—the agrarian community, the glue that held the 18th century together, is replaced by cyberspace." For Gingrich, the information age inspires a new form of communal self-governance. He envisions high-speed fiber-optic cables running into every home. The result will be a new kind of civic culture, described as an "interactive media culture."

Such "technophiles" of the information age do indeed have much to celebrate. The twentieth century has virtually been the "technological century" (Hughes 1991). The logic of modern science has become one of the driving forces—if not *the* driving force—of modern society. One need only reflect briefly on some of the more obvious cases to appreciate the profound significance technological development has had on modern life. In the realm of transportation, for example, we have gone from covered wagon to space travel in less than seventy years. In medicine we have progressed from vaccinations for typhoid fever and smallpox to heart transplants and genetic engineering. In communications we have jumped from radio and telephone to television and the Internet, the media of the information society. Against such developments, there can be no question that technology has had a dramatically profound impact on the quality of twentieth-century life.

At the same time, however, the concept of the "technological century" serves as much as an ideology—a technocratic ideology—as a description of contemporary society (Leiss 1990). We can examine this technocratic ideology through its most contemporary manifestation, the information society.

Information Society as Technocratic Ideology

Ideologies function to blur and conceal important distinctions. The idea of the information society, as ideology, serves to conflate technological advance with social progress (Winner 1977). By failing, for example, to sort out the differences between the kinds of welfare benefits resulting from computer-assisted medical diagnosis and the warfare potentials of computer-guided missile systems, the concept obscures the need to examine basic social choices embedded in technological development. The information technologies, Drucker and Toffler would have us believe, simply rule through self-evident beneficial effects. Rarely do high-tech hopefuls question the social priorities to which such technologies are devoted. As Archer (1990, 108) puts it, "user friendly technology has thus become like health, obviously a good thing."

Moreover, the information society ideology conflates the concepts of information and knowledge (Lyotard 1986). Missing from the concept is the recognition that information obtains meaning only through intellectual processing or manipulation. Without interpretation, the data carried by the increasing flows of information are as meaningless as they are overwhelming. Where the information ideologues see the increase of information making everyone smarter, others see the emergence of a society divided by those with and those without expertise (Beck 1992). Indeed, as the experts get more information, larger and larger numbers of people seem less and less involved in complex issues that confront modern social life.

Against this picture, it is difficult to imagine how power in the information society might be returned to the people. As things stand, it is hard to see this new society giving birth to a self-realizing community of citizen inquirers. More likely are larger numbers of people who report little interest in, or concern for, a technological world that they increasingly find personally meaningless. Such an assessment is backed by numerous trends: great numbers of people simply drop out; others turn to forms of New Age philosophy and spiritualism (Ross 1992). Indeed, newspaper surveys show that a vast number of their readers turn first to their horoscopes, with many reading little else. For most of these people, the wonders of the technological age are largely experienced as new forms of entertainment and leisure, whether faster cars, video recorders, computer games, or techno music (Archer 1990).

Other critics see the information society creating an increasingly global technocratic world. Rather than small self-actualizing communities, they see a world increasingly linked by computerized networks that leave little room for meaningful citizen participation. Indeed, for many critics, nothing is more troublesome than the implications of these trends for democratic governance (Fischer 1990).[2] In sharp contrast to the pluralist politics that have long defined American democratic practices, they see this new societal formulation as portending a much tighter system of interlocking economic and political institutions. At the level of both the national state and the global enterprise, the technological and organizational commitments of this system are said to generate goals and problems (economic priorities, technical exigencies, and political necessities) that frequently result in the curtailment of policy options and choices. Governance, according to these writers, increasingly devolves into a consideration of what is "feasible" given the constraints of the institutional arrangements (Ellul 1964; Habermas 1970b; Winner 1986). The process opens the door to increasingly sophisticated forms of expertocracy that offer fewer and fewer opportunities for meaningful public deliberation (Fischer 1990).

None of this is to say that there is no politics in the information society. Rather, it is to argue that they are a peculiar form of "apolitical" politics predominantly concerned with instrumental adjustments to basic techno-industrial trends. Such a politics is largely concerned with the rationalization of social institutions and practices to better conform to and facilitate the logic of the technological juggernaut. In this view, basic social choices are said to be limited by the technological "imperatives" of the social system. In the process, ideals of equality and democracy are replaced by the standards of meritocracy and efficiency.

This brings us to the final problem. The prophets of the information society conflate morality with instrumental rationality (Feenberg 1991). For these writers, morality is reduced to a mere by-product of the powerful constraints of technological change. Technology, they argue, provides the means that enable humans to achieve their desires. Thus, as an instrumental value, technology exists to serve our needs and desires. As changes in interests and desires occur, technology only requires an amoral, objective process of evaluation and adjustment. The act of desiring something, however, is not enough to qualify it as "good." Moreover, the wants of some people need not be beneficial for humankind as a whole (Stanley 1978). To judge technological changes as a unilinear process of general human advancement can serve only to legitimate the power of those who set the priorities, regardless of the costs imposed by particular technologies on other parts of the population.

Historical experience, especially twentieth-century history, is littered with the tragic consequences of such thinking. From Auschwitz and Hiroshima to Chernobyl and the environmental crisis, we gain an acute sensitivity to the ambiguities of this conventional understanding of technological progress. Such disasters make clear that the task of defining and legitimating technologies as socially "progressive" or "good" requires more than an instrumental rationality (Winner 1986). As a social value judgment, "good" can only be judged against the criteria of moral discourse.

Thus, in this critical view, the information society identifies a world in which questions of social choice lose out to the instrumental thrust of the techno-economic system. It is a society in which basic moral and political questions have given way to the cold logic of science and technology. Archer (1990) argues that we now inhabit a world in which the claims of positivist science dominate both our public and private lives, with the boundaries between the two dissolving as information technologies increasingly invade the home. Moreover, high-tech enthusiasts seldom critically assess the possibility of reversing this further instrumentalization of everyday life. Few, for instance, question why the computer seems only to have sped up the work process rather than to have made it easier, as promised. Science and technology rule to the exclusion of other modes of thought and without major opposition. The ultimate direction unquestioned, we rush forward with a full head of steam.

Technocratic Reason

This technocratic thrust is rooted much more in a way of thinking than in a specific set of political activities. All too frequently, writers have dismissed the technocratic thesis on the grounds that experts, although increasingly powerful, remain subordinate to top-level eco-

nomic and political elites (Brint 1994). Although experts largely remain subordinate, the argument overlooks the less visible discursive
politics of technocratic expertise. Not only does the argument fail to
appreciate the way this technical, instrumental mode of inquiry has
come to shape our thinking about public problems, but it neglects
the ways these modes of thought have become implicitly embedded in
our institutional discourses and practices. The argument, moreover,
misses the fact that such technical languages work both directly and
indirectly to hinder the participation of ordinary citizens, as it underplays—if not denigrates—everyday moral vocabularies. Technocratic
reasoning has its origins in the unique configuration of ideas and
practices that emerged in the West during the past several hundred
years. Traced back to the eighteenth-century French Enlightenment,
particularly to the writings of Henri Saint-Simon (1964) and August
Comte (1830), technocratic thought is fundamentally founded on an
unswerving belief in the power of the rational mind's ability to take
control of the natural and social worlds. In epistemological terms, it is
a "rationalist" orientation grounded in the principles of positivism, or
today "neopositivism." It relies, as such, on empirical measurement,
analytical precision, and a concept of "system," which provides the
foundation of a worldview. As a methodological calculus, it constitutes a practical, instrumental, logical, and disciplined approach to
solving problems and achieving goals (Bell 1971).

Today it is not uncommon to hear—especially from empiricists—
that "positivism," in whatever form, is dead, and to raise this issue is
only to beat down a straw man. In fact, however, neopositivist epistemology is still very much alive in many disciplines, economics being
the most important example in the social sciences.[3] Beyond economics, positivism is most clearly recognized in sociology and political
science in the abstract deductive models of rational choice theory,
itself derived largely from economic theory (Cohn 1999). Even more
important, the principles of neopositivism are still basic to many of
the practices of our institutions, including cost-benefit analysis, forms
of strategic planning, statistical monitoring of outcomes, and peer
review of review panels (Hajer 1995). And not least, it is very much
alive in the curricula of the social sciences. In disciplines such as political science and sociology, for example, experimental hypothesis testing and statistical analysis are still presented as the ideals, although

the vast majority of their members seldom — if ever — practice these methods. One manifestation of this contradiction is that students often ask why they must first master these concepts and practices when — like most of their professors — they plan to ignore them, turning instead to qualitative methods.

Basic to neopositivism and its rationalistic worldview has been an ambitious (if not arrogant) epistemological assumption: that the positivist method is the *only* valid means of obtaining "true knowledge" (Fay 1975). Still today, some say that modern neopositivism will in time subordinate all other modes of thought to its principles. Moreover, supporters believe that rigorous adherence to the methodology will eventually pay off in the discovery of valid empirical regularities, if not the "laws" of society. Not only is such knowledge said to make possible the resolution of many of our economic and social problems but also to facilitate the rational design of social systems in ways that enable us better to predict and manage, if not altogether eliminate, the persistent conflicts and crises that plague modern society (Hofferbert 1990).

In concrete terms, then, neopositivist theory gives shape to an abstract and technical formulation of society and its problems. Social problems, conceptualized in technical terms, are freed from the cultural, psychological, and linguistic contexts that constitute the lens of social tradition. Breaking the "recipes of tradition" and "ordinary knowledge" through the power of this unique abstract language, the neopositivist form of thought creates an illusion of cultural and historical transcendence, which in turn sustains a sense of political, cultural, and moral neutrality (Bowers 1982). In pursuit of the most efficient problem-solving strategies, typically expressed in the precise but abstract symbols of mathematics, experts appear to objectively transcend partisan interests. Their technical methodologies and modes of decision making are said to be "value neutral."

Basic to this process of abstraction is the translation of "experience into theory." The technocratic "pattern of thought understands the phenomenological world in terms of component parts that allow for abstracting the part from the whole, as well as increasingly specialized knowledge of each component part" (Bowers 1982, 531). The result of this "logic of componential thinking" is a view of the world as system (social as well as physical) that can be technically redesigned in ways that make it more efficient and controllable. Critical to this way of thinking is the tendency to see technical solutions as applicable to most social and cultural situations. Problem solving, in short, is reduced to a technical matter of plugging solutions in to different social contexts.

This supposedly neutral, technical understanding of social action is manifested in an administrative conceptualization of problem solving and policy formation. Basic to the objectives of such a managerial strategy is the goal of moving as many political and social decisions as possible into the realm of administrative decision making, where they can be redefined and processed in technical terms. Controversial economic and social problems are thus interpreted as issues in need of improved administrative solutions. The design of these solutions is to be found through the application of managerial techniques (Platt 1969). Policy science has evolved as one of those techniques.

Numerous writers have identified a subtle, apolitical form of authoritarianism in this technocratic strategy. When such expert solutions are legitimated as rational, efficient, and enlightened, it is not so easy for their unwilling recipients to resist their applications. Because of the fundamental differences in the legitimacy and power of their respective languages — technical versus everyday language — the interaction between the technocratic planners and the members of the local community tends to give shape to an unequal communicative relationship, or what Habermas (1970a) has described as "distorted communication." In the policy science literature, this is at times explicitly reflected in a denigration of political decision making, democratic decision making in particular. Terms such as "pressures," "expedients adjustments," or "haphazard acts — unresponsive to a planned analysis of the needs of efficient decision design" are derogatorily employed to describe pluralistic decision making. Such characterizations capture a belief in the superiority of scientific policy methods over political decision processes. If politics doesn't fit into the methodological scheme, then politics is the problem. Some argue that the political system itself must be changed to better accommodate policy expertise (Heineman et al. 1990).

Underlying the technocratic approach is a basic positivist principle that mandates a rigorous separation of facts and values, the principle of the "fact-value dichotomy" (Bernstein 1976; Proctor 1991). According to this principle, empirical research is to be conducted without reference to normative concepts or implications. The effort to do this, however, reflects one of the oldest methodological disputes in philosophy and the social sciences. Pointing to the inherently normative, value-laden character of social and political phenomena, political theorists and normatively oriented sociologists have long complained that the positivists' attempt to separate facts and values reflects a profound misunderstanding of the inherent link between social action and social values.

The critics of the fact-value separation understand society in a very different way than do the positivists. For these interpretive theorists, the social world is not to be understood as a mere set of physical objects to be measured. It is an "organized universe of meanings" that normatively construct the "social world" itself. Such meanings shape the very way ordinary people experience and interpret the world in which they live. They shape as well the very questions that social scientists choose to ask about society, not to mention the instruments they select to pursue their questions. The problem with positivist social science, as interpretive theorists explain, is its failure to adequately take these meanings into account. Rather than drawing on the social actor's own meanings and purposes, positivists tend to construct explanatory models that implicitly *impute* assumptions and value judgments to them. The result, not surprisingly, is bad explanation.

In the process, positivism fails to recognize or acknowledge that social action is invariably oriented toward a conception of the good or desirable (Fischer 1980, 38). To explain such social phenomena adequately, according to positivism's critics, the investigator must get *inside* the situation and "understand" the meaning of the social phenomena from the actor's own goals, values, and point of view. This requires a turn to qualitative approaches better suited to investigate this dimension of social action. What is needed is a methodological framework more appropriately designed to bring together, rather than separate out, the unique and essential aspect of human behavior, the intermixing of the empirical and the normative. Moreover, as we shall see, it is just this interaction between the empirical "is" and the normative "ought" that constitutes both the political and epistemological arguments for democratic participation.

Locating Technocracy

Knowledge Elites in the Decision Process

If technocratic politics manifests itself more as a mode of reason than a traditional form of political action, how then should we identify and understand the role of the professional-managerial strata in the political-economic system? Most critics of the theory of technocracy have dismissed the idea that experts govern on the grounds that they still remain subordinate to the economic and political elites for which they work. Experts, they argue, are not in fact on top; therefore the technocracy argument fails the test of empirical reality. Although it is true that the experts basically remain employees of those with power, this argument falls short on two counts: it underplays the role of knowledge elites in the evolving postindustrial information society, and it misses the more fundamental role of knowledge and expert disciplines in modern society.

First to the question of the role of the knowledge elites. Even in terms of the traditional discussion of technocracy, most of the critics of the thesis fail to appreciate fully the continuing ascent of many high-level experts. That they are not on top does not mean that such experts do not at times make basic policy decisions. Research shows that experts participate more and more with traditional elites in important decision-making functions. Examination of formal organizational structures, both private and public, reveals important changes in the ways decisions are actually made. In the case of the public sector, scholars have identified the evolution of a fundamentally different kind of policy-making process in the modern bureaucratic state (Skocpol 1985; Heclo 1974, 1976; Hall 1993; Keren 1995). The traditional roles of political parties and politicians, researchers find, have increasingly given way to administratively based policy experts. The study of a range of policy domains shows that policy decisions, at least during specific periods, are better understood as the outcomes of evolving "learning processes" among experts within governmental institutions than as the struggles of external political forces. Technically trained administrative and policy experts at times even determine the direction and development of policy (Fischer 1990). This is especially the case in economic policy but can also be seen in areas such as education and administrative policy.

Even a critic of the technocracy thesis such as Brint (1994) acknowledges the importance of experts in economic policy but curiously relegates the concession to a footnote. As he puts it, "Expert-dominated policy making on significant issues does sometimes occur in *politicized* settings, but, outside the economic policy domain, these are rare events and they depend on at least one powerful political leader (usually the president or the governor) and the disinclination of others to mount a contest" (Brint 1994, 249). First, economic policy is the biggest and most significant policy domain in Western industrial state systems. Why should it be downplayed?

Second, that a political leader implicitly supports and sanctions expert decision making should not blind us to the real power that experts have in such cases. Even if they were to be fully in charge in societies that still call themselves democracies — which they are not — such knowledge elites would still need the cover of political legitimation, something that could be provided only by a powerful political leader. One would especially expect this to be the case in "politicized" policy issues. Moreover, we should in no way discount the *non*politicized settings that have in some cases simply been left to the experts. This in itself is surely a sign of technocratic advance. That politicians and the public have sanctioned expert decision making in areas that are no longer controversial policy issues does not diminishes the power of the policy professionals. In many cases, it is in fact a sign of the acceptance of their superior skill and competence — some might even say "legitimacy" — to determine policy.

No one, for example, would argue that the dominant role of physical scientists in the shaping of nuclear regulatory policy suggests the relative unimportance of this domain. As an area that has faced both periods of quiet consensus and heated political controversy, nuclear regulatory issues have been and continue to be dominated by physicists. While it may be true that physicists have proved at times to disagree among themselves on the role of nuclear energy, policy agreements are still largely played out in scientific terms, most often among the scientists themselves. Seldom during periods of such controversy have the politicians simply stepped in and taken the decisions away from the scientific community. Even though different politicians have at various times sided with, and adopted the decisions of, one group of scientists over another, the policy debates are almost invariably carried out in terms that privilege those who possess technical knowledge.

Third, the argument that such expert policy making can happen only when others are disinclined to mount a contest does not always square with the facts. President Reagan's introduction of cost-benefit analysis is a good case in point. There was serious opposition to the introduction of cost-benefit analysis as the primary test for new policy regulations, but the measure not only prevailed as a presidential executive order but remains in effect today despite continued opposition (Noble 1987). By all measures, the executive order establishes nothing less than a technocratic standard par excellence as policy decision criterion. To argue that such examples do not count is to fail to recognize technocracy as a phenomenon in process. That it has not yet fully emerged should not keep us from seeing its steady ascent. That technocrats do not govern should in no way hinder us from acknowledging their increasingly powerful decision-making roles.

This growing influence of experts can, moreover, be seen in the ascent of "policy communities." Today each policy area is the domain of a professional group made up of a network of policy experts, entrepreneurs, administrators, researchers, and writers who specialize in the particular area, for example, health, welfare, environment, and transportation (Heclo 1978; Kingdon 1995). Aptly described as "hidden hierarchies," such policy communities have a disproportionate influence not only over the definitions of specific policy issues but over decisions regarding both the advisability and the feasibility of various solutions. Although such communities still have to sell their ideas to the political elites, it is not infrequent for the basic ideas that emerge from them to become public policy (Schneider and Ingram 1997). Indeed, in face of the escalating complexities of advanced technological societies, it is appropriate to assume that both the ideologies and practices of expertise will continue to expand, often at the expense of traditional elites and assuredly at the expense of the broader public. Although politicians still serve to legitimate such ideas, that they are forged and shaped in expert communities can and does confer substantial power on policy professionals.

With respect to the political process more generally, it is no longer enough for its leaders to rely on the strategic exercise of their political influence. For those seeking to extend their political influence, the complexity of modern policy issues necessitates attention to policy arguments. Regardless of their political strength, interest groups and social movements without access to expertise can scarcely participate in the policy process, certainly not effectively. One sign of this has been the great proliferation of think tanks and the politics of oppositional expertise or counterexpertise, a theme to which we return (Fischer 1991a). The lack of access to policy-relevant knowledge hinders the possibilities of an active and meaningful involvement in the political decision processes for the large majority of the public. Similarly, others have identified the appearance of "policy discourse coalitions" that form among experts and political leaders (Hajer 1995). Such discourse coalitions formulate and advance policy strategies very differently than do traditional party coalitions. Even while party elites retain their *formal* authority, they must increasingly justify their decisions by appeal to the technical analyses of their coalition experts.

With regard to the public, it becomes increasingly clear that in many policy domains, politics more and more becomes a struggle between those who have expertise and those who do not. This is especially the case in technically based fields such as environmental policy. Indeed, access to technical knowledge and skill has allowed those with the power to legitimate their political decisions in these areas. Conversely, as we shall see in later chapters, the lack of access to such knowledge hinders the possibility of an active and meaningful involvement on the part of the large majority of the public. One of the most important contemporary functions of technocratic politics, it can be argued, rests not as much on its *ascent* to power (in the traditional sense of the term) as on the fact that its growing influence shields the elites from political pressure from below (Laird 1990). Not only are experts socially situated between the elites and the public, but their technical languages provide an intimidating barrier for lay citizens seeking to express their disagreements in the language of everyday life. Speaking the language of science, as well as the jargon of particular policy communities, becomes an essential credential for participation. Indeed, in the cases of highly developed professions such as medicine and law, the credential is formally conferred and regulated by the state.

The experts, then, are not in political control, but their information and methods become key resources in the governance of modern society. Not only does access to technical knowledge and skill sustain the power of the top-level political and economic elites, but the *languages* of economics and social science generally now have a profound influence on the shaping of political and policy discourse.

Technocracy as Metapolitics

Disciplinary Power

The traditional state-centered theory of power has hindered our ability to recognize the discursively based expert powers dispersed throughout the social system. The focus on the political position of the technocrat in the decision structure misses the more fundamental power of professional discourses. Indeed, Foucault has demonstrated the ways that this emphasis on structure and position has blinded us to the more subtle but profound nature of professional power. Recognition that the most significant power of the professional is lodged in basic conceptual categories of thought and language opens the door to a discursive understanding of the role of the expert disciplines in modern society.

For Foucault (1972) and his followers, the "political" in the contemporary world can no longer be understood adequately in terms of dominant elites and centers of power. Power now has to be seen as multiple and diversely decentralized. Whereas politics in modern political and social theory is largely understood in terms of state power and law, mainly designed to impede or promote the action of individual citizens, from Foucault's point of view, power is dispersed throughout the spectrum of social relations. Manifested in multiple, ubiquitous forms, political power no longer just belongs to the state alone: it is in effect everywhere. It is at work among psychiatrists who determine the social and medical status of homosexuality, the street-level social workers who interpret the categories of poverty, or the judges who decide the obligations of the father toward his family. Indeed, the very deception of modern politics, according to Foucault, is found in the pretense of confining the location of power to the central government, filtering from awareness its many, ever-present forms (Foucault 1973, 1980). From a poststructural perspective, these newer forms of discursive power are basic to the professional disciplines themselves. As the agents of expert discourses, the professions constitute the techniques and practices that disperse power and social control away from the formal centers of governance.

Centers of power, to be sure, have not disappeared. It would clearly be a mistake to disregard the role of state power, which Foucault referred to as "juridical power." Rather, the point is that an analysis that overemphasizes the centers of power in modern society is altogether insufficient. Foucault's critique of modernity is anchored to his analysis of the rise of extensive administrative forms of regulation — what he calls the "disciplines." From the seventeenth and eighteenth centuries onward, the emerging professional disciplines increasingly took charge of the complex processes by which individuals are made into objects of study — defined as social objects in need of organization and regulation. Such concern — Foucault would say "obsession" — with rational control epitomizes the goals of modernity. Modernity, in Foucault's analysis, can be understood as freeing individuals from the constraints of the Old Regime in order to subject them, for their own good, to the new disciplinary authorities of factories, jails, schools, hospitals, and state administrators. Forging together knowledge, profit, and power, the spread of the new disciplinary order provided a way of controlling large numbers of people, rendering their behavior stable and predictable, without using uneconomical and ostentatious displays of sovereign power, in particular military or police force, which can risk open rebellion on the part of the masses.

Expert disciplines thus took shape at the intersections of words and things, power and knowledge (Foucault 1973). Their regulatory discourses produce "truth," in the sense that they supply systematic procedures for the generation, regulation, and circulation of statements. The knowledge produced is a part of the discursive practices by which rules are constructed, objects and subjects are defined, and events for study are identified and constituted. Such disciplines function in such a way that they can be massively, almost totally appropriated by certain institutions (prisons and armies) or used for precise ends in others (hospitals and schools). At the same time, they remain irreducible to — and unidentifiable with — any particular institutional form or power in society. Rather, disciplines invest or colonize modern institutions, linking them together, honing their efficiency, extending their hold.

Foucault's most concrete illustration of this "power/knowledge" regulatory relationship is his now famous analysis of modern penal institutions (Foucault 1977, 1980). As a "technology of power," modern prison discipline and control works through an interaction of legal rules and procedures, scientific knowledge of criminal behavior, architectural design for maximum surveillance, and administrative processes. Organized as technique, power in this form cannot be located in any one subject — or subjects — managing and guiding the prison system. Instead, power is lodged in the discursive systems and practices that make up the institutional complex. In the process, the prisoner is transformed through the operations of the particular sets of discursive practices that now have an existence of their own. That is, prisoners, as subjects, are rationally reconstituted. Particular prison wardens come and go, but the discourses and practices of criminal justice have a life of their own. Such power is not located in any particular subject; nor is there any single agent to rebel against. Moreover, recognition of these disciplinary mechanisms casts a quite different light on the belief that modern criminology represents a humane advance over earlier barbarisms. Although criminology has left behind the physical dimensions of torture, they are replaced with more subtle but nonetheless stressful manifestations of psychological manipulation and control.

Professional disciplines, operating outside of (but in conjunction with) the state, are thus seen to predefine the very worlds that they have made the objects of their studies (Sheridan 1980). Because this power is *exercised* rather than *possessed* per se, it is not the privilege of a dominant elite class actively deploying it against a passive, dominated class. Disciplinary power in this sense does not exist in the sense of class power. Instead, it exists in an infinitely complex network of "micropowers" that permeate virtually all aspects of social life. For this reason, modern power cannot be overthrown and acquired once and for all by the destruction of institutions and the seizure of the state apparatuses. Such power is "multiple" and "ubiquitous"; the struggle against it must be localized resistance designed to combat interventions into specific sites of civil society. Because such power is organized as a network rather than a collection of isolated points, each localized struggle induces effects on the entire network. Struggles cannot be totalized; there can be no single, centralized, hierarchical organization capable of seizing a single, centralized power. For this reason, argues Foucault, resistance can only be leveled against the horizontal links between one point of struggle and another (Foucault 1984).

Foucault's analysis raises a host of fascinating questions for a reconsideration of the technocracy debate. Clearly, one is the location or constitution of the "political." If power is everywhere, it can legitimately be investigated at the margins as well as the center of society. Recognition of power at the margins permits us to shift our attention away from the central state politicians to other actors such as social movements. Indeed, poststructuralists engage this perspective to explain the unexpected outbreaks of protest movements in the 1960s and 1970s. Occurring outside the established social and political arenas, feminist, antinuclear, gay, environmental, and other citizen movements have increasingly turned to culturally oriented politics concerned more about the social and political *questions* being asked than the *answers* per se. The politics of these new social movements offer, in short, a new metacritique of existing institutions and practices.

If the critical challenges to a technocratic system are to be found in the eruption of transgressive social acts at the margins of society, the poststructural or postmodern critique of expertise implies the need for displacing the established discourses with "local knowledges," which in many cases they only replaced. In this world of multiple realities, professionals lose their unquestionable claim to superior rationality. One of the primary tasks of a reconstructed concept of professional practice becomes that of authorizing space for critical discourse among competing knowledges, both theoretical and local, formal and informal. The task, as Foucault sees it, is to foster critical argumentation in absence of a privileged discourse (Foucault 1980, 1984).

Unfortunately Foucault said very little about how these processes of critical discourse and local resistance might actually arise. Indeed, a major critique of Foucault's work has centered on just this point. His poststructural theory and method leave unclear the role of agency in this process. Only in his last years did he begin to acknowledge this and, toward this end, introduced his concept of local resistance. But the relation of this concept to his analysis of disciplinary power was never worked out before his death. The task here can be understood as taking up this rediscovery of local knowledge and its potential contribution to reconstructing professional discourses, although in ways that differ from a poststructural approach.[4]

Conclusion

Eighty years ago John Dewey asked how citizens could participate in political decision making dependent on knowledge experts. Since then, the question has only grown in importance. What was then a forward-looking philosophical polemic has emerged today as one of the most pressing questions of contemporary democratic theory.

Of course, there is nothing about the contemporary recognition of the problem that has made it less controversial or contentious. From Drucker to Foucault, modern-day understandings of the relationships among technology, expertise, and democratic participation are seen to be anything but consensual. Indeed, contemporary disagreements about scientific technologies and democratic governance are basic to contemporary ideological debate. Both right- and left-wing populists hold out a return to grassroots democracy as the key to revitalizing American society. But the arguments of both groups, as seen here, differ dramatically when it comes to the potential impacts of the information society in fostering a more participatory democracy. Unfortunately both sides respectively argue as if information's connections to either democracy or political control are obvious and relatively automatic. Missing from these discussions is a serious examination of the nature of information and the conditions under which citizens might be able to use it. And it is just here that this work enters the discussion. The following chapters argue that a more participatory grassroots society is a realistic possibility but caution that there is nothing inevitable or straightforward about bringing such a society about. In the pages that follow, I argue that in a complex technological society, such an outcome can only be the result of serious and sustained efforts to think through the intellectual and institutional connections that connect knowledge to democracy. In particular, I seek to rethink the role that experts might play in making these interconnections possible.

2. Professional Knowledge and Citizen Participation: *Rethinking Expertise*

Give people some significant power and they will quickly
appreciate the need for knowledge, but foist knowledge upon
them without giving them responsibility and they will display only
indifference. — Benjamin Barber

Over the past thirty years, the authority of the professional expert,
defined as someone with mastery over a body of knowledge and its
relevant techniques, has become the source of growing concern in
Western societies (Haskell 1984; Brint 1994; Dyson 1993).[1] At times,
this concern has been expressed directly in the form of public protests
against the arrogance and elitism of professional encroachments into
public and private life. Such demonstrations have occurred against
scientists performing chemical tests on unknowing citizens, risk assessors pronouncing the safety of a hazardous waste incinerator in a
highly populated neighborhood, doctors performing (or not performing) abortions, regulators withholding experimental treatments for
AIDS, biotechnologists genetically altering and irradiating foods,
medical researchers performing unnecessary experiments on animals,
physicians denying their patients' requests to die, educators busing
children to distant neighborhoods, psychologists and social workers
telling families how to raise their kids, among others. The majority of
such protests have roots in a kind of "techno-pessimism" that is now
widely found in Western societies. A *Newsweek* survey in 1995, for

example, found that perhaps as much as a quarter of the adult population in the United States is skeptical about the increasing embrace of high technology and the professional emphasis on "technological fixes" to pressing societal problems. As Sale (1995, 785) puts it:

> These neo-luddites are more numerous today than one might assume, techno-pessimists without the power and access of the techno-optimists but still with a not-insignificant voice, shelves of books and documents and reports, and increasingly numbers of followers. . . . They are to be found on the radical and direct-action side of environmentalism, particularly in the American West; they are on the dissenting edges of academic economics and ecology departments, generally the no-growth school; they are everywhere in Indian Country throughout the Americas, representing a traditional biocentricism against the anthropocentric norm; they are activists fighting against nuclear power, irradiated food, clear-cutting, animal experiments, toxic wastes and the killing of whales, among the many aspects of the high-tech onslaught.

Although open protests have tended to occur only sporadically, polls show a steady decline in the public's confidence in, and respect for, professions and their technologies. Rather than a group of experts dedicated to the public good, professionals are widely perceived as a group more interested in increasing their own authority, power, and wealth. Berube (1996, 15) has captured the public's "violent ambivalence" about professionalism:

> On the one hand, Americans know to get a professional if they want a job done right; on the other hand, they know that professionals shroud themselves behind mysterious organizational affiliations and incomprehensible technical jargon. On the one hand, professionals are expert, reliable, accredited, trustworthy, brave, loyal, honest, and so on. But on the other, professionals are arrogant, exclusive, self-serving, money-grubbing careerists — and they purchase their status by discrediting everybody else as "amateurs."

Whereas such public criticisms have tended to vent a generalized anger, critical social scientists and radical professionals have often expressed them more systematically (Fischer 1990). Both the professional-managerial class's allegiances to the top elites and the subtler micropolitics of their practices have been widely discussed among critical social scientists. Especially important, during the past two decades there has been a slow but steady growth of protest movements within the professions themselves. The elitist, ideological, and manipulative tendencies of such experts and their methodologies have been a major focus of conferences, intellectual debates, and media discussions, including the so-called "science wars" debates in the second half of the 1990s (Kanigel 1988; Ross 1996). Radical critics frequently use phrases such as "tyranny of expertise" and "conspiracy against society" (Illich 1989; Lieberman 1972). Critics not only accuse experts of failing to generate solutions relevant to the diverse range of interests in society as a whole but also charge them with using their professional authority and methods to buffer power elites against political challenges from below.

Professional experts, in short, have been portrayed as perpetuating the social injustices plaguing modern Western societies (McNeil 1987; Illich 1989; Wineman 1984; Sassower 1997). Their severest critics see them as having misappropriated their social status and specialized knowledge to serve both their own interests and those of elites intent on maintaining their dominance over the rest of society. And rather than involving a matter of professional malfeasance, the charges relate to the expert's more subtle but fundamental role in the larger social system. By virtue of the professional's middle-level position in the societal hierarchy — that is, between management and labor, government and citizen — he or she typically tends to adopt the system's own definitions of its problems. That is, because most professionals receive their rewards (largesse, status, and authority) from those above them, they commonly come to see the world through the eyes of the elites who license and employ them (Larson 1977). The definitions professionals adopt and build into their practices are thus generally conditioned by elite opinion. The practices they minister are infused with an elite interest in stable social control, if not political domination (Hoffman 1989). As such, experts, through their power to define the client's problems (whether those of an individual or a whole community), often impose definitions and meanings that speak at least as much to the system's imperatives as to the client's needs. This is the essence of expert's mediating relationship between elites and the mass citizenry.

In the professional disciplines, such charges have not been taken lightly. Alternative or movement-oriented professionals have commonly spoken of a "crisis of the professions" (Withhorn 1984). By and large, critics attribute the failures to the professions' overly technical and hierarchical conceptions of theory and practice. Such disciplinary strife has typically centered on charges of normative and epistemological limitations, giving rise to the demand for social and political relevance. Most commonly, this has taken the form of calls for value-oriented, humanistic, and critical approaches to theory and practice (Fischer and Forester 1987). This has become the foundation for a slow but gradually emerging set of alternative professional practices that seek a greater role for lay participation in the research process. Before turning directly to these alternative practices, however, it is important to clarify the question of participation more generally, in particular as it pertains to expertise. Why should we, for instance, think that citizens and experts can — or even want to — participate jointly in expert inquiring processes? Why should we expect citizens to be able to participate meaningfully in research and decision making concerning complex technical and social questions?

Why Citizen Participation?

Public participation, as understood here, is about deliberation on the pressing issues of concern to those affected by the decisions at issue. Or as Bohman and Rehg (1996, 1997) puts it, deliberation is the normative rationale for participation. In this section, we explore such citizen participation in two parts: the first concerning the citizens' abilities to participate, and the second the experts' need for citizen participation. With regard to the former, citizens, it is argued here, are much more capable of grappling with complex problems than generally assumed. While there is, to be sure, a great deal of ignorance in the general public, many citizens are much more intelligent than the Mencken-like stereotypes generally suggest. With respect to the experts, their advice and conduct have seldom been as enlightening and virtuous as their professional ideologies would have us believe. In fact, they have supplied few unassailable answers to our pressing problems. At times their solutions have even turned problems into much bigger ones. In the realm of public policy, their failures often rest as much or more on normative neglect as on the thinness of their empirical analyses. Pulling together the implications of these two arguments, the discussion contends that the solutions to many complex problems are found through more rather than less interaction between citizens and experts. I will argue that far from being just a concern of democratic political theory, such interaction can be necessary in solving many pressing social problems. The question is, What form might that interaction take?

Citizen participation, as a basic political value, is a slippery concept to describe or judge (Nagel 1987). At least on some level, almost everybody is for it; on another, many are quite skeptical of its value in practice. Given the current state of public opinion and citizen participation in the United States, many ask, is it not in fact better to rely on the experts? Is not expert decision making more likely to support the values of social justice (the liberal argument) or efficiency (the conservative argument)? In view of nationalistic, class, and racist attitudes found in segments of the general populace, are we not wiser to invest our trust in the knowledge elites?

Those who have problems with participatory solutions typically argue that the elite professionals should govern in the interest of competence. The experts are the ones with the knowledge and skills needed to render the competent decisions required for effective social guidance. It is an argument typically buttressed with ample survey data showing low levels of knowledge, interest, and participation on the part of the public. For many, the public emerges as something to worry about, if not fear.

Some seek to bolster this position with the argument that there is already too much participation in Western political systems and that much of it is little more than a reflection of the public's limited, self-serving understanding of the complicated problems confronting the country. But this argument refers to interest group politics rather than genuine citizen participation. Interest group politics are not to be misconstrued with citizen involvement in the sense at issue here. Although they speak in the name of large numbers of people, such groups are typically run by a small core of people at the top of their organizations. Indeed, interest group politics has seldom proven to be participatory democracy in action. This, in fact, has become a controversial issue in contemporary environmental politics. Many grassroots environmentalists in the United States, especially those identified with the environmental justice movement, strongly complain that the big environmental Washington-oriented organizations have lost touch with the local citizenry. Having become caught up in so-called "Beltway" politics, such organizations increasingly represent their followers only on paper. It is a political pattern reproduced in one issue area after another.

Others will point to the failures of government participatory schemes as evidence of citizen disinterest. Here too the evidence is far from conclusive. If people have neither interest nor faith in government agencies, as is the case with large numbers of American citizens, why should we expect them to rush to participate in meetings sponsored and orchestrated by the government? Should it even turn out to be in their own interest, little in the experience of most citizens makes this unmistakably clear. Is it not just as reasonable to believe — sometimes correctly — that such participation is only a window dressing for decisions that will be made by others? In fact, how much evidence is there to the contrary?

What can we really expect from the public? To get at an answer, one has to first cut through a good deal of political rhetoric. As we saw, the position that experts should make the decisions is as old as the technocracy movement that followed the Enlightenment (Fischer 1990). Throughout the nineteenth century, various forms of this argument have been basic to the advance of the professions. Modern-day conservatives, moreover, have typically supported this view as a variant of elite politics. On the other side of the issue, leftists of all stripes all too often hold out radical participation as the alternative to technocratic elitism (Chomsky 1989; Morrison 1995). In this view, much of the expert discourse about public policy is mystification. There is little reason to believe, it is argued, that citizens cannot meaningfully participate in the deliberation of the issues. But both of these arguments, those of the Right and the Left, are grounded as much in ideologies as in practical political experiences. In an effort to avoid the excesses of these warring ideologies, I shall take the position that while no evidence suggests that the general citizenry can altogether reject the experts and go it alone in a complex society, the citizenry is more intelligent than many conservative politicians and opinion researchers suggest. I further contend that although citizens need experts, the experts — especially policy experts — themselves need citizen assistance much more than their professional ideologies have acknowledged.

Public Ignorance or Lack of Opportunity?

The question of public ignorance is more complicated than is usually recognized or conceded. Part of the problem is the lumping together of a wide range of citizens under the category of "the public." Averaged together, opinion surveys can produce quite depressing portraits of the public's intellectual capabilities. But there is good reason to believe that significant portions of the citizenry are more intelligent than usually credited. For this matter, we have to recognize that most experts are themselves only members of the public when it comes to other areas of expertise. Given the extreme complexity of many fields, an expert in one field rarely has expertise in others. The rocket scientist might be the contemporary symbol of high-level expertise, but he or she is seldom qualified to perform medical surgery.

What is more, a lack of interest in politics is often mistaken for public ignorance. This misperception jumps over a more fundamental question: namely, whether the public is inherently incompetent to engage intelligently in political matters, or whether its low level of activity only reflects the populace's limited opportunities to develop the interests and participatory skills required to engage meaningfully in public issues. In response to those who fear public ignorance, one can — as many have — just as easily point to the failure of the political systems to socialize its citizens for an active role. In the case of the United States, for example, Americans are first and foremost socialized into the role of consumer rather than citizen. (And here, I think most would admit, Americans don't seem as ignorant or passive as they do in the public sphere — certainly an argument against the innate stupidity of the populace.) Missing from Western political systems are well-developed political arrangements that provide citizens with multiple and varied participatory opportunities to deliberate basic political issues. In the case of the United States, for example, a Kettering Foundation study found that citizens are ready and willing to express this concern themselves. Extensive interviews across ten U.S. cities registered that citizens are not only acutely aware of their remoteness to the political system but are eager for constructive involvement in public life.[2] They express resentment toward those who write them off as ignorant and apathetic and are angry about being "pushed out" of the process by party politicians and lobbyists. One citizen poignantly captured the feeling in the following words: "I'm never aware of an opportunity to go somewhere and express my opinion and have someone hear what I have to say." In the words of one commentator, the study shows that "ordinary citizens seem to share the democratic theorists' concern that our democracy does not offer structures for citizen deliberation and involvement in public decisions" (Hudson 1995, 136).

Although it is difficult to pinpoint the blame in such a complicated matter, one surely has to single out the failure of political parties, especially in the United States. The traditional function of a party is to convey the public's opinions to the political decision makers. But U.S. parties, in this traditional sense, have collapsed and serve only as electoral shells; they have become little more than political labels behind which well-financed candidates organize their electoral bids. The parties' traditional role as communicator has been taken over by the public media. But the media, financed by commercial advertisers, fails as well. By mixing the selling of products with the task of informing the populace, the media emerges as a highly compromised instrument of elite/mass society.

To lament this disappointing state of affairs — widely acknowledged throughout the country — is not necessarily to call for direct or radical democracy, as some seem to quickly assume. Too often the call for more participation is posed as a fundamental challenge to representative democracy. Such arguments misconstrue participation with direct or radical participatory democracy. One can easily acknowledge that citizens in a complex society cannot be — or are incapable of being — involved in all decisions at all times and still call for more participation, especially given the disturbingly low levels of citizen engagement in American society. To do so is not to entertain the kind of utopian fairy tale that radical democrats are typically accused of promoting. Given that only a small percentage of the U.S. public is actively engaged in politics, the effort to increase participation hardly need be considered extreme.

This work concedes straightaway that even though direct democracy has a role to play in particular cases, it is not a tenable model for the society as a whole. I argue instead for a vigorous participatory democracy capable of supporting representative democracy. Given the elite/mass structure of American politics, representative democracy becomes a hollow elitist conception of democracy. Plenty of political experience in the United States and other Western industrial nations attests to the pitfalls of such a system. Elite liberal representatives have their own interests, and no matter how hard they try to speak for the poor and the working classes, their advocacy remains an elite understanding of what these other groups believe and strive for. History shows, moreover, that a struggle for social justice unsustained by broad participatory support tends to be a short-lived activity.

If representative government is to be worthy of democratic legitimacy, it has to be undergirded by a vibrant local system of citizen participation.[3] Beyond simple platitudes about the need for increased voter participation, representative democracy requires structures and organizations that offer citizens an opportunity to deliberate more directly in the decisions affecting their own lives. Insofar as citizen participation is the touchstone of a democratic system, a society with low participation rates need to be concerned about its status as a democracy. In contrast to those who perceive democracy to be destructive or counterproductive, I argue that it is time to explore the boundaries of the possible, a topic I deal with at length in the following chapters. The work proceeds from the belief that a great deal more participation is both possible and necessary than presently exists in Western democratic systems and that, among other things, this means rethinking the relationship between experts and citizens.

Politicizing Expertise

From Advocacy to Participation

The first step toward a more democratic, collaborative relationship between experts and citizens has been "advocacy research." Advocacy research is a practice put forward by activist social scientists and other professionals aligned with progressive political issues (the War on Poverty, environmental crisis, antinuclear struggles, and the women's movement, among others). As a methodology, advocacy research is pitted against the practices of democratic elitism.[4] In the name of political empowerment, it has taken up the interests of the unrepresented and powerless client. Designed to directly confront the elitist biases of mainstream research, advocacy research has been advanced to facilitate democratic empowerment (Foster 1980; Davidson 1965).

In epistemological terms, advocacy research represents an attempt to transcend the "value-neutral" ideology of expertise by explicitly anchoring research to the interests of particular interest groups and to the processes of political and policy argumentation in society generally. In doing so, it seeks to offset the discipline's allegiances to the dominant political and economic elites, especially as they are manifested in a mediating role between elite requirements and mass demands. By making expertise available to groups otherwise excluded from decision processes, advocacy calls attention to the implicit, hidden, and elitist politics embedded in conventional professional practices. For this reason, it clearly constitutes an important political and methodological step toward a less elitist, more democratic expert-client relationship.

Advocacy has been an important step in the right direction, but it nonetheless failed to fulfill the promise of a genuinely participatory methodology (Kennedy 1982; Kraushaar 1985). In the course of their struggles, many activists have come to recognize that advocacy research is useful for representing views not otherwise heard in the political process but is not well designed for the fundamental requirement of participatory democracy, namely, helping people speak for themselves. In the case of the alternative medicine and urban planning movements, for example, Hoffman (1989) found that the local groups that these professionals sought to represent often ended up feeling that the positions advanced by the experts did not really represent those of the group.

With regard to the issue of poverty, research documented that agency workers often did little to determine if they were fighting for the issues that really bothered the poor. The agency workers' role was essentially program or issue oriented rather than client oriented per se. Their work tended to emphasize making policy changes deemed appropriate by professional knowledge and standards, and influencing the behavior and attitudes of public officials, particularly those seen as lacking expertise. As Kennedy (1982, 34) explained, "They had a genuine desire to assist the poor, but on their own terms, using their methods, and their issues." The resolution of the policy issue, rather than being the development of an ongoing community process, became the primary goal (Gaylin et al. 1978).

Advocacy research's problem thus lodges in the failure to deal with the hierarchical character of both expertise and democratic elitism. As became apparent to many activists in the middle 1970s and early 1980s, advocacy research has merely been grafted onto a system of hierarchical interest group organizations that tend themselves to be oligarchic. The leaders of these organizations are seldom as representative of their constituencies as they purport to be. In many cases, as research shows, the representation of a group by experts leads to an elitism that impedes the possibility of authentic membership participation (Elgin 1984; Michels 1915). Indeed, in recent years, postmodernists have elevated local knowledge and the failures of representation, both political and conceptual, to a basic tenet of their critique of modern institutions (Baudrillard 1983).

Advocacy also failed to recognize that a more democratic policy expertise would also involve a reconsideration of many of the widely used research practices. By and large, advocacy research has employed fairly standard research methodologies; they have simply been directed at different political questions and problems. Few recognized the biases embedded in the conduct of mainstream research itself. Central to the findings of the more critical sociology of science that began to emerge at about the same time is the recognition that many of the hierarchical practices at issue were implicitly supported and facilitated by standard scientific practices. Postmodernists, moreover, have persuasively argued that the sciences have served to define their subjects in ways that work to adjust them to the control strategies of the institutions in question (Foucault 1972). In short, beyond a more progressive political commitment, the democratization of expertise requires new methodological orientations as well. The fundamental argument to emerge from these critiques of expertise is put forth in the following terms: if citizens are to participate in the development of the policy decisions that affect their own lives, the standard practitioner-client model must give way to a more democratic relationship between them.

This does not mean that citizens can themselves simply replace experts, although some modern-day utopians occasionally suggest it. More formally, this view has been described as a "participant-dominated" model of expertise (Rossini and Porter 1985). Even though it might be possible to demonstrate specific circumstances under which such an approach can work, as a general model, it is problematic. Given the complexity of society, the participant-dominated model is clearly beyond reach. More plausible is a "collaborative" or "participatory" model of expertise. Although it is common for mainstream professionals to portray the search for nontechnocratic alternatives as impractical and ill suited for modern technological society, there is growing evidence that paints a different picture. Numerous experiments make clear that citizens and experts can strike a much more democratic balance between knowledge and participation. Many of the experiments have come from the "new social movements," the environmental movement being one of the most important.

Participatory inquiry has evolved in the context of struggles against environmental hazards in both the community and the workplace.[5] As such, it is founded on the efforts of citizens to broaden their access to the information, with a view to research that meets people's own needs (Merrifeld 1989; Fals-Borda and Rahman 1991). Grounded in a critique of professional expertise, participatory inquiry attempts to gear expert practices to the requirements of democratic empowerment. Rather than providing technical answers designed to bring political discussions to an end, the task is to assist citizens in their efforts to examine their *own* interests and to make their own decisions (Hirschhorn 1979). Beyond merely providing analytical research and empirical data, the expert acts as a "facilitator" of public learning and empowerment. As a facilitator, he or she becomes an expert in how people learn, clarify, and decide for themselves. The second half of this book will explicate the model of participatory inquiry in the context of specific cases, along with its implications for the reshaping of the citizen-expert relationship. Chapters eight through twelve will work out both its theoretical, epistemological, and practical implications for giving professional advice.

The Expert as "Specialized Citizen"

None of the foregoing is to overlook the fact that technical issues make it difficult for citizens to participate under the best of circumstances. Complexity will continue to ensure the need for professional expertise. But this only brings us to the other side of the problem; namely, that the experts themselves are not without their own difficulties. Although worries about an uninformed citizenry are scarcely misplaced — as far as the argument goes — during the past two decades or more, we have had ample occasion to worry about as well the "best and the brightest" (Halberstam 1993).

Take, for example, the case of the United States. From the Vietnam War, we learned of the arrogance of the policy planners (McNamara 1995). The result was countless unnecessary deaths and a protracted period of social unrest. Social policy during the liberal era of the Great Society often revealed the thinness of sociological understanding, not to mention the political naïveté, of many liberal social scientists. Programs such as school busing and model cities (some would add affirmative action) provide excellent examples of the failures to foresee the unanticipated consequences of expert advice. Even worse, Reaganomics demonstrated how easily experts can self-servingly embrace a blatantly ideological program despite its disastrous fiscal implications for the country as a whole. What we have learned from these and many more examples — or at least should have learned — is that the elites are themselves no guarantee against folly. Not only do the experts have their own professional ideological commitments, often conflicting with the public interest, but they possess no analytical wizardry capable of resolving our pressing societal problems. Expert judgment, we come to recognize, provides few uncontested solutions or answers. At best, policy advice is an informed opinion. As the new sociology and history of science teach, scientific expertise is not what we have long been told it is. While we still need experts, expertise cannot stand alone.

If this argument is true in general, it is particularly true for the field of social policy. Here the expert's authority is much more ambiguous than in technical areas such as nuclear power or space travel. In matters social, normative assumptions and values are as important as technical analysis. No demonstration of efficiency can ever suffice to convince citizens to accept a social program that they don't believe to be good, right, or fair. From the new social studies of science, we learn that science is laden with social value judgments, judgments typically hidden within the steps and phases of the research process. And it is here that the case for citizen involvement starts to become apparent. When it comes to the basic normative assumptions and social understandings that underlay and prestructure policy research itself, the experts can have no privileged status. Although data remain important to normative social choice, they can never be sufficient. In choices about how we want to live together — or how to solve the conflicts that arise in the struggle to do so — the experts are only fellow citizens.

More precisely, what does it mean to say that citizens have a role here? At one level, this can be understood to mean that citizens should comment on and discuss the social implications of expert analyses. In fact, few in a democracy could deny this role. But if the interaction is carried out on the intellectual turf of the expert, as it typically is, citizens will always come up short in such exchanges. Insofar as experts understand or treat the essence of policy to be its technical core, as do most conventional policy analysts, the citizen's input will remain a secondary, inferior contribution to policy deliberation.

How have we come to neglect these normative political questions? Why have we replaced them with narrower — even secondary — technical questions? Technocratic politics, as seen in chapter 1, has its origins in an effort to challenge and replace such discourses. As a worldview hostile to political discourse, technocracy has sought to transcend — or at least circumvent — politics through expert judgments. Asserting the superiority of the scientific method, the technocrat holds political deliberation to be an outmoded — in some cases even "irrational" — way to solve conflicts. To be sure, few contemporary technocrats openly denounce politics. In systems that define themselves as democracies, all feel compelled to at least pay lip service to the concept. The technocratic argument more subtly manifests itself in a call for improving policy deliberation through improved technical inputs. Although there is nothing wrong with improved technical inputs per se, the effect of the argument — at least as it stands — is to emphasize and elevate technical over political discourse. As political discourse comes to be seen as inferior, typically defined as less rigorous, it is gradually but steadily denigrated.

That professional policy analysis has succeeded in elevating the technical over the political is a testament to the contemporary influence of technocratic ideologies. That expert knowledges emphasize the technical and instrumental rather than the social and political dimensions of policy should come as no great surprise in an age of technical experts. Technical knowledge is, after all, what the experts have to offer.

But in the "real world" of public policy there is no such thing as a purely technical decision. To be sure, all policies have a technical component (some being much more technical than others). Nor can there be any doubt about the need for technical information about what works and what doesn't. But none of this should blur the more fundamental fact: policies are first and foremost social and political constructions. As a uniquely normative entity, a policy decision — like social decisions generally — is constructed around sets of normative understandings and the ways of life of which they are part. Although policies are rules introduced to alter, fix, or guide social and political problems, these problems arise in the course of our continual struggle to live together harmoniously.

If public policy is not technical per se, how should we understand it? As a response to social and political problems, policy inherently combines a mix of social and technical factors, none of which can be understood wholly independently of the other. Far more than an empirical discourse about efficient or effective action — the standard technical conception of policy — a complete policy judgment rests on a series of interrelated discourses, each with its own logics and methods (Fischer 1995). Central to these discourses are basic questions about the social construction of the empirical objects to be assessed (e.g., is a "woman" the same thing as a "lady"?), social choices about a policy's implications for a particular way of life (e.g., does the policy promote individual initiative and self-help?), as well as more specific questions about its application to particular social contexts (e.g., do ghetto residents live in a "culture of poverty"?). Indeed, these are the kinds of questions on which policy is built. Although the full range of normative considerations is seldom problematic at once, all are potentially troublesome, especially as competing arguments shift the contours of a deliberation. In a critical evaluation of public policy, such questions take priority over lesser technical questions about the efficiency of the program under discussion. The experts, in the context of these social assumptions, can themselves only answer as citizens, at best as "specialized citizens."[6]

Given the intrigue of the larger societal-level questions, the issues of social context are easily overlooked. But this is unfortunate. To be usable, knowledge has to be applied to a particular situation or context. Even though policy-oriented social science has geared itself—at least epistemologically—to the search for generalizable propositions (i.e., propositions valid across contexts), social context is decisive in the world of action. One might best think of this as a translation problem: how do we translate abstract propositions into particular contexts?

The translation process introduces two further considerations for citizen participation. One is that a context is always a social construct. Any given situational context means different things to different people, including those who are a part of it. For this reason, the social definition of the situation is crucial to the application of policy-analytic findings. Because there can be no decisive empirical definition of a social context, policy development and implementation without the assistance of those living in the particular social setting to which a policy is to be applied can at best be questionable activities. Not only are the intentions and motives of the locals essential to a proper understanding of a situation, but they also typically possess empirical information about the situation unavailable to those outside the context. While such local knowledge cannot in and of itself define the situation, the "facts of the situation" are an important constraint on the range of possible interpretations.

Once the inherent social foundations of a policy are acknowledged, the door is open to an interpretive approach to policy analysis. Even though such a conceptualization of the discipline defies the conventional wisdom, it gets epistemological support from the newer postpositivist theories of science. What the positivists have failed to grasp, along with much of the Western tradition generally, is that scientific discourse is itself a highly interpretive enterprise. Given this interpretive dimension, science loses its privileged claim as superior knowledge. Empirical science need not fold up shop, but in a practical field like public policy, it has to establish a new relationship to the other relevant discourses that bear on policy judgments. Rather than an inferior mode of discourse, political deliberation has to be recognized as a different type of inquiry with different goals and purposes. In the political world of public policy, scientific discourse cannot replace or circumvent the questions of political deliberation. As we shall see in the highly technical field of environmental policy, the usefulness of an applied science depends on its interactive relationship with political deliberation.

A new question thus poses itself. Rather than which discourse is *better*, the question of the relationship among multiple discourses emerges. Instead of questioning the citizen's ability to participate, we must ask, How can we interconnect and coordinate the different but inherently interdependent discourses of citizens and experts?

The argument is not that citizens should involve themselves in the technical issues of science, although this need not be entirely ruled out. As the case of AIDS activism has shown, not only can citizens learn a great deal about science, but their impact can change the research process itself (Epstein 1996). In general, questions of technical accuracy or competence can be dealt with by counterexperts. The primary issue is more a matter of the experts finding ways to relate their technical practices to public discourses. This challenge, as Willard (1996) argues, suggests the need for a new subject matter, or "epistemics," as he calls it. Where traditional policy expertise has focused on advancing and assessing technical solutions, the new subject would investigate the movement and uses of information, the social assumptions embedded in research designs, the specific relationships of information to decision making, the different ways arguments move across different disciplines and discourses, and the interrelationships between discourses and institutions. Most important, it would involve innovating methods needed for coordinating multiple discourses in and across institutions, a topic to which we return in chapter 12.

The professions have almost totally neglected this epistemic translation of their activities. It can be posed as the major challenge to a more relevant mode of professional practice. Participatory democrats within the professions should place the working through of these epistemic interconnections among citizens and experts and their institutional implications at the top of the research agenda. Whether we are talking

45

about large or small numbers of citizens (e.g., a political party or an advisory group), the prospects of democracy in a complex society would seem to depend on it.

The chapters that follow seek to engage this challenge: How might we go about bringing citizens and experts into a democratic, mutually productive relationship? What might it mean epistemologically? What would it involve for both disciplinary practices and institutional decision processes? Toward this end, this work advances a set of participatory practices designed to facilitate collaborative exchanges among citizens and experts, or, as identified here, "participatory inquiry."

Conclusion

This chapter initiated the task of rethinking the relationship between citizens and experts. The discussion opened with the critique of expertise and an examination of the alternative response within the professions, advocacy research. Although it is a positive step forward, advocacy research falls short of genuine citizen participation. The solution is to be found in more participatory forms of deliberation. The remainder of the chapter examined the underlying question "Why more citizen participation?" and offered reasons as to why policy experts themselves need to seek out closer relationships to citizens. The discussion first showed that there is nothing obvious about either one of these questions; and second, that there is much more room here for innovative possibilities than normally recognized. Against these considerations, the chapter closed by arguing that both public problem solving and democratic governance, especially when they apply to value-laden policy issues, would be better served if the technically oriented, top-down expert-client relationship were replaced by a more professionally modest but politically appropriate understanding of the expert as "specialized citizen."

Although the discussion of expertise here applies to public policy making generally, nowhere is it more important than in environmental politics. The next chapter, for this reason, turns to an examination of the nature of the political and epistemic conflicts between citizens and experts in environmental policy issues.

3. Environmental Crisis and the Technocratic Challenge: *Expertise in the Risk Society*

In March 1986, a nine-page article about the Chernobyl nuclear installation appeared in the English-language edition of *Soviet Life*, under the heading of "Total Safety." Only a month later . . . the world's worst nuclear accident — thus far — occurred at the plant.
— James Bellini

Nowhere are the conflicts between citizens and experts more salient than in environmental politics. All of the concerns that we have raised are central to the environmental issue; one would be unable to understand it independently of the question of technological progress and the role of scientific expertise. Indeed, many take this close relationship between environmental politics, science, and technology to render meaningless the search for nontechnocratic alternatives; democratic alternatives are seen as impractical and ill suited for a modern technological society. This work, however, turns to numerous experiments that make clear the possibility of establishing a more democratic balance between citizens and experts. Many of these efforts, devoted to reconstructing expertise, have come from the "new social movements," the environmental movement being one of the most important.

From an environmental perspective, no one has more sharply raised these issues than the German sociologist Ulrich Beck. We can thus use Beck's work, in particular his concept of the "risk society," to set the stage for the examination of the conflicts over environmental risk, both technical and social, that work their way through the remainder of the book. For Beck, we enter into a new era of late modernity in which the question of what constitutes expertise, as well as who has it and who does not, increasingly represents one of the most basic fault lines of contemporary society.

The concept of the "risk society," formulated by Beck (1986) in the mid-1980s, represents a unique and important contribution to a critical understanding of science and expertise in environmental politics. In an age when most social scientists have specialized in ever narrower domains of social and political inquiry, Beck has attempted to grasp the environmental crisis in its social totality. As an exercise in social diagnostics, he has labored to come to grips with the implications of the crisis for social and political change in Western societies generally. Beck, in this respect, has sought nothing less than to reconceptualize the whole of modern society as a "risk society." In what has proved in Europe to be a most provocative thesis, Beck's work has emerged as the basis of a wide-ranging environmental discussion, including the relation of science to the public. Reaching far beyond the walls of academia, it has been debated extensively in the German public as well. For these reasons, the risk society provides an especially interesting theoretical backdrop against which to situate an analysis of the relationship of citizens to experts in environmental struggles.

The Risk Society

Modern Risks as Social Decisions

To be sure, concerns about danger and safety are not new to human society. Plagues, famines, and other natural disasters have not only posed serious threats to human life throughout human history but also at times even endangered and destroyed entire civilizations. In earlier periods, people mainly blamed the gods, even though people were themselves often much more implicated in the causes than they realized. The difference today is that our mega-technological dangers are unmistakably human made. They are, in short, the result of institutional decisions geared to economic opportunities. They are the utilitarian by-products of techno-industrial strategies. Such decisions, basic to both the peacetime and the military economies, originate in the centers of rationality and prosperity. More directly stated, businesses and states are responsible for them. As the interwoven products of the atomic, chemical, and genetic revolutions, these newer risks distinguish themselves from traditional dangers in other ways as well. They are unprecedented in terms of both visibility and scale. In the case of scale, their destructive capabilities are incomparable with those of earlier forms of disasters. Nuclear radiation is thousands of times more deadly than factory smoke. The cumulative impact of modern toxins on the human body and the wider ecosystem, largely unknown, is after a certain point of exposure irreversible. Moreover, such risks often respect no temporal boundaries. Across generations, they can accumulate in both intensity and complexity.

By "risk society," Beck refers to an epoch in which the dark sides of progress increasingly come to dominate social and political debate (Beck 1992). Essentially, the risk society brings forth that which few care to see and no one wants: the self-endangering, devastating industrial destruction of nature. In a short period of time, this environmental question has, to use Beck's (1995a, 2) words, positioned itself as a potent "motive force of history."

Most basic to Beck's theory is the argument that we can identify a fundamental shift in industrial society. Whereas industrial society traditionally celebrated the production of material goods, since the end of World War II, it has increasingly confronted worries about the production of risks (Beck 1992). More and more people have come to recognize that the technological risks involved in the production of many of these goods have risen to such a level that they become more troublesome than the traditional risks associated with material scarcity. In Beck's risk society, the logic of the production of risks increasingly overshadows the production of goods. That is, the *positive* industrial logic of distributing wealth and social goods is offset in the risk society by a *negative* logic of risk production and avoidance, or what Beck describes as the distribution of social and personal bads. It is not that the earlier industrial society produced no risks but rather that the nature of contemporary risks tend to become more visible and worrisome as a more affluent society demands a better quality of life. Although Beck tends to exaggerate the situation, he is certainly right to argue that technological risk has become a central anxiety of the

time. Given the relative satisfaction of basic needs in industrial societies, such risk avoidance has emerged as a central political issue of
our time.

Accordingly, the politics of socially generated risks introduce new
lines of political conflict. In industrial societies, the historically defining conflicts emerged from class divisions; they concerned the ways in
which the distribution of wealth (and thus the risk of poverty) was
divided among different social groups. Whereas in industrial societies,
social class groupings mainly experienced risk differently, in the modern risk society, their experiences have in many ways started to converge. In contrast to the rich, who largely lived in areas and worked
under conditions that exposed them to little danger, the industrial
working class was traditionally exposed to dangers and threats.

A steel factory, for example, belched out toxic emissions that jeopar-
dized the health of its workers and those who lived near the factory.
Those with money and privilege, however, could simply escape; they
moved away from the source of the risk. But once certain types of risks
and dangers began to exceed their traditional spacial and temporal
limitations — once they were no longer confined to particular communities or social groups — economic means could no longer provide a
secure escape route. The atomic age threats posed by nuclear radiation, biotechnology, or the greenhouse effect can potentially threaten
all social groups — rich and poor — at the same time. For example,
toxic accumulations in the food chain owing to the use of dangerous
pesticides can threaten the entire population. In the risk society, it
matters little whether one lives in the city or the suburbs. Even more
vivid, the radiation poisoning of a nuclear meltdown can reach the
rich as well as the poor, the Southern as well as the Northern Hemi-
sphere. Threats of nuclear weapons, moreover, have held out nothing
less than the possibility of self-annihilation. Against these new realities, Beck finds the quest for safety to now overlay the more traditional concerns about class and distribution.[1]

With regard to visibility, these newer risks are virtually undetectable
without scientific investigation. Not only are the impacts of such risks
in no way obviously tied to their points of origin, but their transmission is often invisible to normal perception, the invisibility of radiation being the classic example. Unlike many other political issues,

environmental risks must actively be brought to the citizen's awareness to be identified as a social threat. Given the highly technical and invisible nature of these risks, the politics of risk intrinsically emerge as a politics of knowledge, typically contested through expertise and counterexpertise. Because the existence of such risks — let alone their origins and consequences — must be deduced by active causal interpretation, they exist in the social world only insofar as there is scientific awareness of them. At every stage in our understanding of such risks, the mobilization of scientific knowledge is central to their description and assessment. This elevates the expertise and status of the knowledge professions to a prime political position in the discourse of risk, leaving little or no room for the layperson. The result is a growing tension between those with and those without knowledge. Indeed, for Beck et al. (1994), this is the central fault line of the risk society, or what in later works he calls "reflexive modernity."

Given the subjective side of the interpretation of modern risks, it becomes increasingly clear that the environmental crisis is as much a crisis of the institutions that have to interpret and regulate risks as it is a physical phenomenon pertaining to natural processes (Fischer and Hajer 1999). Against the scope and character of the environmental problem, as Beck (1995a, b) makes clear, the conventional political institutions — the representative bodies, regulatory agencies, and scientific institutions of industrial society — that are assigned the responsibility of negotiating our understandings of the risks we face clearly fall short of the assignment.

Environmental Crisis as Institutional Crisis

Parliamentary democracy and its bureaucracies, shaped by the politics of class and interest, are no longer capable of adequately controlling and legitimating the technological forces unleashed by corporate capitalism. Moreover, the traditional approach to the resolution of class conflicts — namely, the politics of economic growth — no longer suffices. Although the distributional conflicts of industrial society were eased by making the cake bigger — a positive-sum game — once the cake was perceived as poisoned, Beck (1995b, 128–57) argues, the formula has ceased to work its wonders.

Risk societies are thus trapped by an outdated repertoire of political and administrative responses now inappropriate to modern catastrophes. Consequently we face a paradox: at the very time when hazards and catastrophes appear to become most nefarious, they simultaneously slip through the nets of proof, laws of liability, and systems of compensation with which the legal and political systems attempt to capture and remedy them.[2] Consider first the question of compensation. The collective agreements that have emerged over the past two hundred years for dealing with industrially produced risks and uncertainties begin to buckle under the legal implications of modern megatechnologies. When insurance has to deal with the possibility of destruction across the planet, the pillars supporting the calculus of risk are eroded, if not abolished. Often portending irreparable damage that cannot be spatially and temporally limited, such accidents emerge as events without beginnings and endings. Moreover, as such consequences become incalculable, statistics turn into a form of obfuscation. Coupled with a lack of institutional accountability, the possibility of determining causality becomes hopelessly complicated. Indeed, against this backdrop of incomprehensibility, the very risks themselves tend to lose their meaning.

Or take the legal responsibility of demonstrating liability. Currently, such responsibility lies with the afflicted parties rather than the potential polluters. A legacy of industrial society's faith in progress, the principle is institutionalized in the legal system's assumption that industrial production will be benign unless demonstrated otherwise. Given, however, that companies are the only actors likely to have a good sense of the risk implications of any given process or production in development, no one else is likely to sort out the environmental implications before pollution has begun. Any attempt to demonstrate harm will occur only after people have been exposed to the damage. The prevailing definitions of risk are thus weighted in favor of the polluter. Not only does the legal system demand proof of *post hoc* toxicity (rather than *pre hoc* nontoxicity or safety), but those who must prove toxicity are inevitably less endowed with the detailed skills and information necessary to make a convincing case.

Even if the risks and dangers of earlier industrial societies could have been sufficiently captured and interpreted with the available models of social causation and risk (itself an empirical question), this is no longer a possibility in the newly emerging risk societies. In the face of such uncertainty, the public's response tends to take the form of what Beck (1995b, 56–57) calls "industrial fatalism." The concept identifies a central paradox in Western societies: namely, that while the public must live with the obvious threats of uncontrolled techno-industrial development, its citizens are unable either to account for the existence of such threats or to accurately identify the culpable individuals. Technological developments, as such, would seem almost to take on a life of their own, with those closest to the dangers often registering their hopelessness through a form of denial. To make matters worse, the political and legal systems designed to redress such injuries and grievances tend — intentionally and unintentionally — to render invisible the social origins and consequences of these risks. Thus, possessing neither the ability to identify those responsible for their anxieties nor the mechanism to redress the sources of the problem, those surrounded by such threats tend to retreat to an old and established defense mechanism: they simply choose not to see or hear about them — that is, industrial fatalism. Pulling these threads together, society is left, as Beck argues, to confront a late-twentieth-century crisis with nineteenth-century institutions and procedures (1995a, 1995b). At best, contemporary political and administrative institutions offer little more than a veneer of confidence, constantly broken by the harsh realities of new accidents.

The Illogic of Science

Society as Laboratory

Nothing is more fundamental to the nature of the crisis than the inability of science and its institutions to speak authoritatively in a time so much in need of information and assessment. Although science is essential to the awareness of most modern risks, there is nothing obvious or straightforward about the central role of its discourses and practices. Indeed, the reliance on science in environmental policy making is fraught with tensions and contradictions (Yearly 1992). Consider first that scientific technologies are themselves a cause of most of these modern risks. Nuclear power and biotechnology are the direct by-products of scientific and technological research. Without science we would not be worried about the dangers of these tech-

nologies. At the same time, however, the identification and consequences of risks must in part be couched in scientific terms. It is impossible to detect and debate the threats posed by toxicity without some
degree of scientific knowledge of the chemical and biological processes involved. Furthermore, in terms of alternative production processes, new products, and cleanup technologies, science is a source of
solutions to these risks. The interrelationship of these processes has
worked to raise questions about the epistemological and cultural status of both science and the conduct of contemporary politics. As science in public affairs becomes more and more a politics shaped by
professional expertise, it has unwillingly opened the door to a closer
public scrutiny that has posed basic questions about the legitimacy of
science.

For Beck, the greatest danger inherent to modern science is that it
has turned society into a laboratory. We now confront, in this respect,
a situation that is as dangerous as it is ironic. Whereas science typically seeks reliable knowledge through laboratory experiments, in the
case of contemporary large-scale technologies, the process has been
reversed. Before scientists can learn about the long-term risks of our
mega-technologies, they have to first build and implement them in the
society at large. As Beck (1995a, 104) writes, "nuclear reactors must
be *built*, artificial biotechnical creatures must be *released into* the
environment, and chemical products must be *put into circulation* for
their properties, safety, and long-term effects to be studied." Knowledge about the safety of a nuclear power station, for example, can be
derived only *after* its construction and operation. In this reversal of
normal scientific procedures, the operation of a facility merges with
the testing of the facility. In this "adventure of technological civilization," the scientific experts have transformed society itself into the
laboratory. As the experiment is exported from the laboratory into the
open air of daily life, basic ethical and epistemological questions arise
about the logic and conduct of research.

From an epistemological perspective, once the controllability of laboratory conditions is lost, the very logic that makes possible a precise
conceptualization of research design and the operationalization of
variables collapses. As Beck (1995a, 105) puts it, checking hypotheses
emerges as a fictive exercise as "the opening of the laboratory boundaries requires one to assume theoretically and practically uncontrolla-

ble influences." But insofar as the violation itself is carried out in the name of scientific advance, those who dare to question this new phenomenon are typically accused of being opponents of scientific progress.

How, then, can scientific and technological research that investigates the very things it invents and implements ever be legitimated in traditional scientific terms, namely, through the disinterested pursuit of the natural world? In Beck's (1995a, 105) words, "Genetic engineers, human geneticists, reactor researchers, practitioners of reproductive medicine . . . become beggars or solicitors for their own cause." As this becomes more and more apparent, citizens begin to raise questions that science cannot answer. Quickly the research establishment comes to recognize that scientific legitimation is not enough. As unsettling as it might be, it becomes difficult for scientists to hide the fact that their research now depends on the public's political consent. In the process, "politics comes before research, and research really and literally becomes politics itself, because it must produce and change something in order to develop its scientific rationality at all" (Beck 1995a, 105).

The result is something of a crisis for the scientific community. Not only does the theoretical question of how science is actually practiced take on new interest in academic circles, but the door is opened to public discussion of what science is up to. For larger and larger numbers of people, the growing influence of science and technology gives rise to public fears and disputes about its privileged status. Laypeople express political uncertainty and hesitation not only about the implementation of such experiments in society but about the very direction of scientific research itself. More and more, citizens and politicians speak of the need to regulate and control science. Although the scientific community cries foul, arguing that this constitutes an intervention in the pursuit of knowledge, the public increasingly comes to see that in matters of political goals and social judgment, the scientists are themselves laypersons. The affected citizens begin to recognize not only that the scientists are ignorant of the consequences of their actions but also that the scientists are interested laypersons in their own scientific projects. Because there is no way of really knowing if their experiments are harmful before they are carried out, coupled with the scientists' concerns that public questions and doubts can shrink research funding, professional bias enters as a pervasive phenomenon. As Beck (1995a, 105) ironically puts it, "The suspected thief passes judgment on a robbery."

Underlying these concerns is yet another epistemological question: can science and society learn from mistakes? According to modern theories of science, such correction is the core of scientific rationality. Popper, for example, posited the "falsification" of errors as the cornerstone of scientific epistemology. For Beck (1995a, 106), however, the new situation raises a new question: "Who will decide, however—how, when and upon what basis—whether a social experiment of production as technology has failed?" If experimental research has to be implemented to be tested, then it has "leveraged away its conditions of falsification." In this case, "all accidents and disruptions—for instance, in nuclear power plants all over the world—are experimental findings in a continuing, perhaps undecidable concrete experiment."

Science has tried to cover or disguise these questions by introducing quantitative risk assessment. But statistical risk assessment has largely failed to comfort an anxious public that, often intuitively sensing the limitations of the findings, withdraws its trust from the risk assessment community. Moreover, many philosophers and social scientists have succeeded in showing that statistical analysis of risks serves—both wittingly and unwittingly—to render less visible the more fundamental social questions embedded in the very *design* of technological research. Ignoring or hiding important social and health impacts, such analysis, as Beck (1995a, 21) points out, functions as a form of moralizing in the name of mathematical objectivity.[3] Such calculations represent, as such, "a kind of bankruptcy declaration of technical rationality." Furthermore, studies show that such questions are answered differently from one country to another, from one culture to another.

Most important, people come to recognize that in a democratic society, such moral questions can no longer remain in the hands of the technologists and engineers. People more and more see that the near monopoly that technological thinking has had on such issues must give way to a more democratic form of deliberation. "Precisely because the investigation of effects and risks presumes their production, others—laypeople, the public sphere, the parliament, and politicians—must also have a say; they must regain the power to make decisions in a society that has gone over to shaping its future through technology" (Beck 1995a, 109). Instead of advancing more technical remedies to solve technical risks — that is, more of the same — people begin to recognize the implications of such strategies for a democratic society. Beck puts it this way:

> Risks can be minimized technically. Anyone who depends on them as the only lever to gain and expand some public say in the techno-scientific adventure puts pressure on himself to consent when the safety concerns are alleviated. Democracy beyond expertocracy . . . begins where debate and decision making are opened about whether we *want* a life under the conditions that are being presented to us even by those technologies that are growing steadily safer. (1995a, 109)

The solution is to be found in the development of a more participatory form of democracy, which Beck designates as an ecological democracy.[4] Once society has become a laboratory — and the citizens objects of the experiment — the door morally and politically opens to the public voice. In this situation, discovering truth becomes both public and "polyvocal." Insofar as there are now no genuine experts in matters of risk, the traditional technocratic, monopolistic concept of science has to give way to a more "reflexive" or self-critical concept of science. In Beck's words:

> If the engineers have the say here *de facto*, then it is important to open the committees and the circles of experts and evaluators to the pluralism of the disciplines, extra-disciplinary modes of judgment, and shared decision making that have already been speaking out for some time and have begun to organize themselves. (1995a, 109)

Citizens and Experts

Ecological Democracy as Reflexive Modernity

Immanent in the dynamics of modern risk politics, according to Beck (1986; 1992), is a set of forces that compels society in a more reflexive direction. As the Green critique of industrial growth gives way to a more social understanding of progress, a new form of institutionalized self-criticism emerges that provides individuals with more opportunities to deliberate and recalibrate the regulative principles of industrial society, not just the specific policies associated with it. A simultaneous and interconnected politicization of science and knowledge, coupled with a remoralization of politics, lead to what Beck sees as a new self-critical ecological democracy (1995a). Instead of abandoning modernity, as postmodernists would have us do, Beck sees its yet unfulfilled democratization in the making, a new era of "reflexive modernity."

New social movements are the leading edge bringing this political reflexivity to the fore. Political reflexivity, or "that alarming of the system that will occur as people become aware of the general threats to life in the milieu of a bureaucratically administered security," will depend on the ability of social movements to exploit "the social explosiveness" of modern hazards (Beck 1995b, 2). That public opinion polls in Germany have ranked ecology as "most urgent" is the result of citizen activity that recognizes the threat of the institutional normalization of the potential dangers of self-destruction (1995a, 3). Responding to these risks, both actual and imagined, the environmental movement emerges as an alternative to the parties and interest groups of the conventional political system, whose structures, means, goals, and interests no longer reflect the social experiences of everyday life. Fueled by the emergence of new risks and threats, the environmental social movement's deep-seated challenges to techno-industrial advance usher in the reflexive risk society from the decaying body of an antiquated industrial society. For Beck the "risk society" is a society that can reflect on the risks it faces. It is not just a society that faces risks. Thus, the environmental movements are responsible for bringing in this new critical, reflexive dimension.

There need be nothing obvious or inevitable about the emergence of ecological democracy. Indeed, the standard scientific approach to risks and safety — risk assessment — threatens to transfer more public authority and responsibility to an environmental technocracy and its ever-encroaching technologies. But the question of how we want to live together remains a matter of discursive struggle. What is needed to rescue and vitalize this struggle in a technological society is a self-critical or reflexive practice of science. Moreover, rather than merely trying to ameliorate the unforeseen consequences of new technologies, social decision making about technologies must be introduced at an earlier stage (Beck 1995b). Only by assigning decisions "on technologies to public and political processes before and during the genesis of hazards, can we return the fate of hazard civilization to the realm of [human] action and decision-making" (Beck 1995a, 110).

Beck opens the door here to a more democratic restructuring of science and technology, but unfortunately he doesn't really take us through it. Up to now he has yet to say much about what such a critical or reflexive science would look like. Indeed, his concept of science remains surprisingly traditional. Even though he recognizes science's ability to deal with its normative crisis, he more or less leaves the practice of science itself intact. Although he sees the struggle among those possessing knowledge and those lacking it as the fundamental fault line of modern politics, he never questions the concept of knowledge itself. Concentrating his focus on a call to open up to laypeople and politicians the forums and committees in which science is deliberated, Beck never really questions the conventional understanding of science. In the end, the issue seems only to be which scientists should we believe, rather than a deeper critique of science. What we are left with is the need to look for new ways to further democratize the processes of counterexpertise.

Where Beck's work tends to converge on this question of knowledge, he largely puts it off as a discussion for another time or context. Insofar as his entire concept of reflexive modernity depends on a new kind of science, this is a mistake. For one thing, there is nothing obvious about the ability of the public to participate in expert decisions. One should not merely assume the inevitability or feasibility of public participation. The issue is, as we have argued, is one of the crucial challenges confronting the future of democracy in a complex technological society. Whereas participatory democrats tend merely to assume people's abilities to participate in all decisions at all times, conservatives warn of a crisis created by too much participation on the part of an uninformed public, the result of which is described as "systems overload," if not simply bad decisions (A. King 1975). What is worse, modern political experience offers little guidance in this situation. Indeed, one can easily argue that high levels of public ignorance and low levels of public participation offer the participatory democrat little encouragement (Willard 1996). But this argument, of course, is itself open to the "naturalist fallacy"; namely, the principle that existing behavioral patterns do not preclude the possibility of different behaviors by citizens at a later time.

Fortunately, however, we need not leave the topic here. New developments in the sociology of science, social constructivism in particular, help us to rethink the question of what knowledge is, including how people participate in its construction. Several writers — for example, Wynne (1996) and Fischer (1990) — have taken the next step of opening risk politics and risk assessment up through a postpositivist, constructivist reconceptualization of what we mean by science and knowledge. The remainder of this chapter will outline Wynne's brilliant critique of Beck's understanding of science. The discussion will serve, in this respect, as a foundation for the later introduction of the discussion of the possibility of participatory expertise.

Scientific Expertise in Postpositivist Perspective

The Constructivist Interpretation of Political Reflexivity

For Wynne (1996), Beck's account of a new political consciousness triggered by the growth of unmanageable risks is based on an overly objective or "realist" conception of both environmental risks and the expert knowledges that seek to characterize them. Although experts and their knowledges are central to the kind of societal transformation Beck envisions, he works with an outmoded understanding of these concepts. Never does Beck's risk society thesis really question the meaning of expertise and knowledge, especially the social and cultural bases of their indeterminacies. The political issue revolving around his discussion of the politicization of expertise is more a question of how people decide which experts to believe or trust than a question about the usefulness or appropriateness of the type of knowledge put forward. The possibility that other types of knowledges might more appropriately speak to the worries of an anxious citizenry, or that citizens might themselves have their own forms of knowledge on such matters, gets at best only passing mention. The question of knowledge, however, has itself increasingly emerged as the critical issue underlying the concerns of growing numbers of citizens. As Wynne makes clear, a full explication of the reflexivity of the risk society requires a more *cultural* interpretation of the science that is being politicized. Not only does the effort to elaborate the risk society thesis need to be refounded on a hermeneutic understanding of the processes that might bring it about, but such an understanding is essential to the design of an inquiry process appropriate to a reflexive society. More specifically, the theory of the risk society requires a constructivist interpretation of scientific knowledge and expertise.

First, however, a closer look at the problem. Implicit in Beck's model is the idea that lay citizens are losing trust in a science and expertise they feel has increasingly betrayed them. Behind this anxiety is a sense that our contemporary institutions can no longer manage and control the escalating risks that modern science and technology have unleashed. The traditions of modernity taught us to focus on our capabilities to know and plan, but this newer configuration of circumstances redirects our attention more to the limits of our knowledge, in particular to the unanticipated consequences resulting from the applications of modern technologies. Such uncertainties have shaken the public's faith in the experts. After having long trusted experts generally, citizens are confronted with the task of choosing which experts to believe and trust. Although this interpretation is not wrong, it fails to capture the more subtle, critically important dimensions of the citizen-expert relationship. Beck's portrayal of this interaction, shared as well by Anthony Giddens (1990), is based on an overly instrumental-calculative interpretation of the citizens' cognitive orientation. In this view, citizens, enlightened by counterexperts, actively invest trust in particular experts through deliberate choices between recognized alternatives. Underlying this perspective, as Wynne (1996) makes clear, is an overly rationalistic conception of the citizen-expert relationship. Required is a more cultural and constructivist analysis of the relationship.

Several issues need to be addressed here. The first concerns the assumption that citizens at some earlier time held out an unqualified trust in professional experts. This idea, Wynne argues, largely rests on a confusion of three interrelated concepts: unreflexive trust, private ambivalence, and reflexive dependency. Involved here are several misconceived but mutually reinforcing assumptions. First is the idea that earlier experts enjoyed an ostensibly uncontested public status that, for theorists like Beck, becomes equated with public trust. This, as Wynne explains, reflects a limited understanding of public acquiescence. Rather than simple acceptance of experts, the phenomenon is intricately tied to the citizen's recognition of his or her social and institutional dependency on them. It represents citizens' awareness of the power that experts and expert institutions have over their own lives and the need to assume at least in part a strategic orientation toward this dependency. To draw a closely related analogy, the fact that people often don't tell their bosses that they think they are wrong does not have to mean that they are in agreement with them. Such theories neglect the fact that public ambivalence toward, or alienation from, institutions need not be manifested in overt behavior or expressed commitments (Fischer 1995, 14–15). They misread the lack of observable dissent as a sign of the existence of trust. As Wynne argues, this cannot be assumed; it is itself an empirical question.

Second, there is the issue of the role of risk in the contemporary distrust of experts. Is this distrust, and the possibility it might hold out for a new reflexivity in late modernity, the result of enlightened calculation and deliberate choices made by citizens exposed to new hazards? Although there are surely instrumental-calculative dimensions to citizens' deliberations about the "objective" facts and fears of the risks they confront, social constructivist research increasingly shows that such a perspective misses or understates the more important social and cultural interpretive dimensions that underlie and condition such thought (Harrigan 1995). Indeed, it neglects the very dynamics of the thought processes that lead to a deeper reflection on the institutions and processes of which the experts and their arguments are a part. To get inside the question of how quiescence can reflect social ambivalence and alienation, we need a more sophisticated conceptualization of the cultural processes interconnecting social agency, identity, and dependency. Sociocultural analyses show that under conditions of social dependency, overt signs of acceptance and trust are often better treated as "virtual trust," or "as-if trust" (Wynne 1996, 50). Finding themselves in situations of social dependency, citizens are often compelled to act "as if" they trust the experts, keeping major doubts to themselves.

Taking the argument a step further, this position suggests that the citizen's relationship to the expert has always had a reflexive dimension. In Wynne's (1996, 47) words, "sociological work which has identified the unrecognised sense of dependency and lack of agency which pervades public experience of and relations with expert institutions has also identified the unsuspected reflexive ways in which this is manifested as a lack of overt public dissent or mistrust." It shows "how people informally but incessantly problematize their own relationships with expertise of all kinds, as part of their negotiation of their own social identities." Aware of the extent of their dependency on specific expert institutions, citizens are cognizant of their inabilities to take full charge of their own situation (Wynne 1992; Irwin and Wynne 1995). Neither the fact that no public dissent is manifest nor that citizens are frequently incapable of fully articulating the boundaries of their dependencies should blind us from recognizing that "these lay processes are deeply imbued with reflexivity" (Wynne 1996, 50).

Citizens' responses to scientific expertise are bound up and conditioned by an appreciation of, and accommodation to, social dependency on expert institutions. This, in turn, directs our attention to the expert institutions themselves. To what extent do these institutions socially construct their responsibilities to obscure or mystify their own role in the creation of modern risks and dangers? As Wynne (1996, 51) avers, to what degree are they constructed to appear "as Acts of God which no one could have possibly anticipated or controlled"? Institutions that reconstruct "history so as to confirm their blamelessness whilst attempting to manufacture public trust and legitimation are prima facie likely to be undermining public trust rather than enhancing it."

Such an interpretation resonates with Beck's analysis of the scientific community's denial of its role in the production of modern risks. Science's efforts to conceal its responsibilities for its expert systems unintentionally work to magnify the public's concern. Because citizens are aware of their dependency on these systems for protection against harmful risks, the citizen ambivalence to the experts can serve to enlarge the citizens' own perceptions of the risks. In short, the citizens' deliberate choices involve more than the alternatives presented by the experts. The lack of trust in expertise itself factors into their own calculation of the risks.

The point has been illustrated in sociological research pertaining to risky large-scale technological systems (Fischer 1991b). A closer look at the perceptions of the risks presented by these systems, as such research shows, makes clear the need to consider the interplay between physical and institutional factors. That is, technologies, large-scale technologies in particular, pertain to much more than the engineer's understanding of their physical properties. Technology, as such, must be reconceptualized as "a set of integrated techno-institutional relations embedded in both historical and contemporary social processes" (Fischer 1991b; Wynne 1987). Recognizing a technological system to be an inseparable web of socio-organizational and technical processes is basic to an appreciation of the way people experience and perceive technologies. Social perceptions and everyday evaluations of technological risks — those of workers as well as citizens — are rooted in their concrete social experiences with these organizational decision structures and their historically conditioned relationships. Any single technical "event" or "decision" is in fact located within a continual socio-institutional process, which is itself an integral dimension of a technology, especially a large-scale technological system. For example, sociological evidence shows that workers cannot altogether divorce their response to physical risks from their attitudes toward social relations in the plant, particularly those pertaining to managerial practices. If the workers' social relations with management are pervaded by mistrust and hostility, the ever-present uncertainties of physical risks in the plant are amplified (Fischer 1991b).

Finally, it is widely assumed that citizens take their lead in such matters from the experts. Drawing on a range of careful studies of the interactions between citizens and experts, Wynne shows that this assumption often gets it backward. Of particular importance here are the studies of opposition to nuclear power by Welsh (1993, 1995). Contrary to the conventional wisdom — which holds that such opposition only began in the 1970s, in large part thanks to the environmental movement — Welsh's work demonstrates that long before nuclear experts began disagreeing with one another in public, many lay citizens were actively questioning the expert findings, advanced as authoritative justification for a rapid expansion of the industry. Dissent among the experts is frequently generated and supported by the existence of a public backdrop of doubt and disaffection. That is, at critical moments, dissent in the expert community may well follow the lead of dissent from citizens.

All of these considerations underscore the need for a more complex understanding of the public's relationship to expert systems. Even when citizens do believe and trust in expert bodies, this trust is much more conditional and indeed more fragile than standard interpretations reflect. Laypeople's relationships to expertise are thus more skeptical of, more ambivalent to, and more alienated from expert institutions than is generally recognized. To rectify this, a better understanding is required of the kinds of knowledges and thought processes that lay citizens bring to the task of assessing risks (Irwin 1995).

Here too one can find an emerging body of literature to draw on. Numerous studies of public risk perceptions show that ordinary people bring more to their evaluations of risks than acknowledged by the expert's reductionist framing of citizen responses (Slovic 1992). This research demonstrates more conventional work to have neglected two basic dimensions of risk perception. The first concerns the social context in which risks are embedded: Is the risk imposed by distant or unknown officials? Is it engaged in voluntarily? Is it irreversible? Second, experts make assumptions about the character of the risk situation that are quite removed from the experiences of those at the actual site. For instance, experts typically take for granted the competence and trustworthiness of those controlling the processes in question. But the framing of such competencies and commitments often embodies a model of the social world and the relationships of laypeople that is open to question (Wynne 1996). When taken as tested models, such risk frameworks naively serve to impose prescriptive commitments on problematic situations. What may begin as testable hypothetical assumptions about the social world (for instance, whether nuclear plant personnel always rigorously adhere to the regulations) subtly get transformed into prescriptive norms for social control, rules creating a reality that serve to confirm scientific evaluations. Such assumptions must be recognized as incipient social prescriptions of particular social orders and cultural identities. Although their role is buried in an objectivist discourse, the scientists who impose these models are themselves acting as naive sociologists. Often this can give them the power to create the very social assumptions and implicit commitments that tacitly shape their own knowledge. Thus the very institutions designed to control "direct" physical risks are a crucial aspect of both the risk society and risk research more generally. Insofar as alienating models of social behavior are built into the decision processes of the institutions supposedly advancing solutions to environmental risks, depending on these institutions becomes risky business. Science is discovered to be laden with expropriated social meanings no longer scrutinized and discussed by the social actors to which they pertain. Against this light, it becomes important to rethink the neglect or dismissal of the "ordinary" or "informal" knowledge that laypeople may possess about the validity of the experts's real-world assumptions — for example, about the production, use, or maintenance of a technology. It is here that laypeople must be acknowledged to have a legitimate claim to deliberate such practices and assumptions (Wynne 1996; Fischer 1990, 1995). Moreover, given that laypersons can have a lifetime of experience with the social negotiations embedded in these practices, they can often experientially intuit the thinness — if not ignorance — of an expertise based on a one-off response to an evaluation or consumer survey (Wynne 1996). Such recognition manifests itself as a deep source of suspicion and distrust.

In the name of "propositional truth," then, the objectivist decision methods built into modern expert institutions delete the cultural and moral foundations that make it possible to understand the very institutional processes to which they pertain. Not only do such explanatory models empty the meaning out of the lives of the relevant social actors, but even worse, they refill these lives with their own imposed meanings. As Wynne argues, the conflict at this unarticulated hermeneutic level creates alienation toward, and refutation of, contemporary institutions. Such alienation, in turn, promotes the "cultural politics" advanced by extrainstitutional social movements.

Basic for Wynne are the ways in which "this largely negative hermeneutic dimension" is strongly "amplified by the enhanced role which social science has played in environmental and risk policy work," particularly in rational-choice models of environmental analysis, social surveys of risk acceptability, and social-psychological research on public risk perceptions (Wynne 1996, 60). These methods impose instrumental, individualist, decisionistic, and essentialist models of the human action in the name of "neutral" scientific observation. In Wynne's view, then, we must consider the possibility that the very growth and intensity of citizens' perceptions of risk have resulted in part from the increased intervention and influence of the positivist social sciences in these public realms. Such a science is left with no honest intellectual choice but to relinquish its claims to a value- and meaning-free status, so long proclaimed by the doctrines of positivism, and open itself to a more deliberative understanding of its own procedures and practices.

We turn in the following chapters to a more detailed examination of the effects of these practices and the possibility of building in the neglected local and lay knowledges. Whereas others have identified the citizen's possession of such informal knowledge, the main emphasis here is on how the professions should reorient their own practices to accommodate such knowledge. As the next step toward this end, chapter 4 turns to a more detailed discussion of how positivist science has come to play the central role in environmental policy making.

4. The Return of the Particular

Scientific Inquiry and Local Knowledge in Postpositivist Perspective

It is . . . possible to visualize a kind of social science that would be very different from the one most of us have been practicing: a moral-social science where moral considerations are not repressed or kept apart, but are systematically commingled with analytic argument without guilt feelings over any lack of integration; and where moral considerations need no longer be smuggled in surreptitiously, nor expressed unconsciously, but are displayed openly and disarmingly. Such would be, in part, my dream for a "social science for our grandchildren." — Albert O. Hirschman

In the preceding chapters, I argued that the restructuring of professional expertise must be based on a more reflexive approach to science. Making good on Beck's call for a democratic restructuring of science and expertise, I argued, requires rethinking our understanding of knowledge itself. Moreover, to come to grips with the ways in which citizens respond to expert practices, it is necessary to open up the discussion to a consideration of the social and cultural foundations of knowledge(s). In doing so, as will be seen, it becomes possible to come to grips with the thought processes and forms of knowledge that ordinary citizens employ in their own deliberations about environmental risks, including their assessments of expert opinion. As a first step in this process, this chapter examines the epistemology of such an alternative, a constructivist understanding of science, concentrating in particular on its application to the social science and policy expertise. Toward this end, the chapter spells out a constructivist, discursively oriented "postpositivist" conception of science.[1]

The chapter is divided into three sections. The discussion first takes up the failure of the empirical social sciences to make good on their long-promised predictive theory of society. In significant part, these failures are traced to a "neopositivist" epistemology and its "universalist" perspective on knowledge. Neopositivist social science is seen to cling to an understanding of the physical and natural sciences that is no longer unquestionably accepted in these so-called hard sciences.

In the second section, the analysis focuses on the postpositivist alternative. For the constructivist-oriented postpositivist, the solution is to turn from the traditional emphasis on scientific proof or verification to a contextual, discursive understanding of social inquiry. In this perspective, knowledge is understood to be nested in a context of time and local circumstances. Instead of merely suggesting postpositivism as an alternative epistemological orientation, this section offers this discursive or "argumentative turn" as a better description of what social scientists already do. Finally, drawing these strands together, the third section examines the more concrete implications of a discursively oriented approach for expert policy practices. In particular, this concluding discussion addresses the relationship of postpositivism to empirical inquiry.

The Limits of the General
Social Science and Public Policy

Postpositivism is in large part a response to the failures of the contemporary social sciences (Giddens 1995; Lemert 1995; Wallerstein et al. 1996). Neither have they developed anything vaguely resembling a predictive "science" of society, the original promise, nor have they been able to provide effective solutions to pressing social and economic problems, the later policy-oriented commitment (deLeon 1988; Baumol 1991; Fischer 1998). Scarcely having gone unnoticed in the social science community itself, a number of policy scholars have devoted considerable thought to the question of what might constitute "usable knowledge" (Lindblom and Cohen 1979; Innes 1990; Fischer 1995). Thus far, the mainstream effort has not been impressive.

This is not to say that the social sciences have had no impact on public issues. To the contrary, the influence of the social sciences is everywhere to be found in contemporary political discourse. But the role has been more to *stimulate* the political processes of policy deliberation than to provide answers or solutions to the problems facing modern societies. Although such deliberation is generally acknowledged to be important to effective policy development, this "enlightenment function" is not the analytic mission the policy-oriented sciences have set out for themselves (Weiss 1990). More ambitiously, they have sought to develop universal methods and practices designed to *settle* rather than stimulate debates. This traditional "neopositivist" understanding of the policy-analytic role not only rests on an epistemological misunderstanding of the relation of knowledge to politics; its continued reliance on a narrowly empirical mode of inquiry also hinders the field's ability to more directly approach what it can — and should — do, namely, to improve the quality of policy argumentation in public deliberation. The field's outdated epistemological orientation impedes its ability to develop methods and approaches that facilitate this important enlightenment-oriented discursive function. Postpositivist policy inquiry, as we shall see, is in significant part an effort to rescue this policy-analytic mission by setting it out on its own epistemological footing.

What is "neopositivism"?[2] Emerging first as an epistemology to explain the methods of the physical and natural sciences, neopositivism (or "logical empiricism") has also supplied the ideals of contemporary social and policy science (Hawkesworth 1988). As such, it has supported the rise of a social science in pursuit of quantitatively replicable causal generalizations, public choice theory being the most rigorous contemporary variant. Most easily recognized as the stuff of the research methodology textbook, neopositivist principles emphasize empirical research designs, the use of sampling techniques and data-gathering procedures, the measurement of outcomes, and the development of causal models with predictive power (Miller 1991; A. Kaplan 1998). In the field of policy analysis, it is manifested in quasi-experimental research designs, multiple regression analysis, survey research, input-output studies, cost-benefit analysis, operations research, mathematical simulation models, and systems analysis (Sylvia et al. 1991).

The only reliable approach to knowledge accumulation, according to this epistemology, is empirical falsification through objective hypothesis testing of rigorously formulated causal generalizations (Popper 1959; Sabatier and Jenkins-Smith 1993, 231; Hofferbert 1990). The goal is to generate a body of empirical generalizations capable of explaining behavior across social and historical contexts, whether communities, societies, or cultures, independently of specific times, places, and circumstances. Not only are such propositions essential to social and political explanation, but they are seen to make possible effective solutions to social problems. Such universal propositions are said to supply the cornerstones of theoretical progress.

In the face of limited empirical successes, neopositivists have had to give some ground. Although they continue to stress the rigorous quantitative pursuit of general principles, they have retreated from the more ambitious project. Today their goal is more to aim for propositions that are at least theoretically generalizable at some future point. An argument propped up by the promise of computer advances better able to amass and correlate data, it serves at minimum to keep the original epistemology intact. But the argument misses the point, as postpositivists are quick to point out. The problem is more fundamentally rooted in the neopositivist social scientists's misunderstanding of the nature of the social. It is a misunderstanding lodged in the very conception of the universal, value-free objectivity they seek to reaffirm and extend.[3]

Constructing Knowledge

Situational Context and Social Assumptions

Contemporary postpositivism is rooted in both the natural sciences and the history and sociology of science. With the advent of quantum mechanics and chaos theory in physics and evolutionary theory in the biological sciences, growing numbers of scientists have come to reject a static view of the universe in favor of one in flux (Toulmin 1990). From this research, we come to see that the traditional understanding of reality that has guided the physical sciences — the very conceptualization on which the social sciences have tried to base themselves — is itself a contested topic in the "hard sciences." Indeed, we learn that what one observes in the physical world depends in important ways on where one stands, the very problem the social sciences have long tried to escape by adapting the methods of physics to the social world.[4]

Moreover, on the heels of these discoveries arrived new historical and sociological observations about the nature of scientific behavior. From these "postempiricist" studies, we have learned that both the origins and practices of modern science are rooted as much in social and historical considerations as they are in the disinterested pursuit of truth per se.[5] Such investigations, particularly those of "social constructivist" sociologists, have shown the activities of empirical inquiry — from observation and hypothesis formation through data collection and modes of explanation — to be influenced or shaped by the theoretical assumptions of the sociocultural practices in which they are manifested (Rouse 1987). In this view, science is itself a form of human action.

From such investigations, we see the degree to which the application of scientific methods to particular problems involves social and practical judgment. The model form of the experiment, for example, proves to be more than a matter of applying a causal research design to a given reality. As often as not, as Latour (1987) has shown, reality is discovered to be fitted to the empirical instrument. In some cases, science gets its results by identifying and organizing those parts of reality that are amenable to the research design. In other cases, it goes beyond such selection processes to restructure the social context (Rouse 1987). These critical investigations make clear that a proper assessment of research results has to go beyond empirical data to examine the practical judgments that shape both the instrument and the object. Although such judgments structure and guide the research process, they are almost never part of the research paper. The formal write-up of the results is organized to conform to the official judgment-free logic of science.

None of this is to imply that science should not be taken seriously. It means, rather, that the thing we call "science" has to be understood as a more subtle contextual interaction between physical and social factors. Whatever constitutes scientific truth at any particular time has to be seen as more than the product of empirically confirmed experiments and tests. Such truths are better described as scientific *opinion* or *belief* based on an amalgam of technical and social judgments. From this perspective, there can be no such thing as a "fact" as the term is conventionally understood. Facts, in the natural as well as the social world, depend on underlying social assumptions and meanings. What is taken to be an objective fact is in effect the decision of a particular community of inquirers and the theoretical presuppositions to which they subscribe.

Nowhere are the implications of this critique more important than in the study of politics and policy. The network of presupposed assumptions underlying social and political propositions reflect particular social arrangements; the assumptions are themselves influenced by politics and power. Not only is one of the basic goals of politics to change an existing reality, but much of what is important in the struggle turns on the sociopolitical determination of the assumptions that define it. Policy politics, as numerous scholars have made clear, are about establishing definitions of, and assigning meaning to, social problems (Edelman 1988; Gusfield 1981; Best 1989). The effort to separate out meaning and values thus cuts the very heart of politics out of social inquiry. In its search for value-neutral generalizations, neopositivist social science detaches itself from the very social contexts that give its findings meaning, a point to which we return later in the discussion.[6]

Seen in this light, the outcomes of such research can at best be relevant only to the particular sociohistorical understandings of reality from which they are abstracted. Moreover, positivism's attempt to empirically fix a given set of social and political arrangements tends to reify a particular reality. By neglecting or diverting attention away from the struggles to challenge and change such arrangements, social science — wittingly or unwittingly — serves as much to provide ideological support for a configuration of power as it does to explain it.

Both the interpretive nature of the social object and the meaning of the empirical findings render neopositivist science an easy target for those who wish to dispute the validity of specific experiments or tests. At best, such research can offer a rigorous and persuasive argument for accepting a claim. But such an argument cannot *prove* the issue. Those who prefer to dispute a claim can easily find problems in the myriad of social and technical interpretations and assumptions em-

73

bedded in both the research design and empirical practices. Nowhere is this more obvious than in the endless confrontations over the validity of environmental science, which has given rise to a full-scale politics of "counterexpertise," as we have seen in the preceding chapter. Working with the same findings, groups on both sides of an issue easily construct their own alternative interpretations of the data. Each side, in name of the "facts," seeks to offer a better social construction of the evidence (Hannigan 1995).

The constructivist view helps us to see that in such policy debates, it is more generally the deeper social and cultural factors, rather than the "facts" of the arguments, that play a decisive role in citizens' assessments of the competing views. By drawing our attention to the sociocultural contexts that underlay the citizen-expert relationship, the constructivist approach shows us the ways in which citizens interpret the "objective" assessments of professional experts within the context of the citizens' own normative cultural experiences and the social dependencies inherent to them. Insofar as these sociocultural factors are inaccessible to neopositivist methods, as we saw in the preceding chapter, such research often tends to underestimate the degree to which laypersons are ambivalent toward, or alienated from, professional experts and their institutions.

From this perspective, an understanding of the social world depends on knowing what social actors believe reality to be. While as Innes (1990, 32) puts it, "this does not require us to accept a shared meaning as the only way to understand something, such meanings are essential 'data' for any analysis." What we call "knowledge" of the social world is the outcome of a negotiation between those with more "expert knowledge" and the actors in the everyday world, including the experts themselves. For this reason, the process of investigation necessarily deeply involves the expert in the normative understandings and processes of everyday life. As such, the process of knowing cannot be understood as the exclusive domain of the expert.

To recognize this deeper interpretive role of the cultural context underlying social research is not to argue that it is never worth carrying out an empirical test. The postpositivist objective is not to reject the scientific project altogether but rather to recognize the need to understand properly what we are doing when we conduct one. Postpositivism, in this respect, is best explained as an attempt to understand and reconstruct that which we are already doing when we engage in scientific inquiry. Recognizing reality to be a social construction, the focus shifts to the circumstantial context and discursive processes that shape the construction. We turn at this point to a closer examination of the epistemology of this alternative approach.

From the General to the Particular

The Postpositivist Alternative

During the second half of the twentieth century, as Toulmin (1990, 186) explains, "the problems that have challenged reflective thinkers on a deep philosophical level . . . are matters of *practice*." Of particular importance, in this respect, have been the problems posed by the threat of nuclear war, medical technologies, and environmental degradation. "None of them," he writes, "can be addressed without bringing to the surface questions about the value of human life, and our responsibility for protecting the world of nature, as well as that of humanity." As a result, the intellectual orientations that have dominated the last three centuries have come under sharp criticism. The long-standing emphasis on "the universal, the general, and the timeless" is being rethought to make room for "the particular, local, and timely." The shift, in short, has been away from the overly narrow formal conception of rationality that has shaped the history of the social sciences.[7]

Different theorists have employed different names to capture this rethinking of science. To avoid terminological debate, I refer here to the movement more generally as "postpositivism." Given that social science takes place in, and refers to, a particular context, this postpositivist orientation recognizes that social science offers an *account* of reality rather than reality itself. This is not to say there are no real and separate objects of inquiry independent of the investigators. Rather, it is to emphasize that the vocabularies and concepts used to know and represent objects are socially constructed by human beings. Scientific accounts are understood to be produced by observers with differing degrees of educational training, research experience, perceptual capacities, and ideational frameworks. The goal of postpositivism is to understand how these varying cognitive elements interact to discursively shape that which we come to take as knowledge. Toward this end, postpositivism's reconstruction of the scientific process is founded on a "coherence" theory of reality emphasizing the finite and temporally bounded character of knowledge (N. Brown 1977; Stockman 1983).

In contrast to neopositivism, coherence theory addresses the undeterminedness of empirical propositions. Seeking to describe a world that is richer and more complex than the theories constructed to explain it, the goal is to capture and incorporate the multiplicity of theoretical perspectives and explanations that bear on a particular event or phenomenon. The task is to bring to bear "the range and scope of interpretive standpoints that have won a place" (Toulmin 1983, 113). Alongside the empirical inquiry, postpositivist coherence theory includes the historical, comparative, philosophical, and phenomenological perspectives. In the process, empiricism loses its privileged claim among modes of inquiry. While it remains an important component of theory construction, it no longer offers the crucial test.

Given the perspectival nature of social and political phenomena categories, knowledge of a social object or phenomenon can better be understood as something that emerges more from a discursive interaction — or dialectical clash — of competing interpretations. Whereas consensus under neopositivism is inductively anchored to the reproduction of empirical tests and statistical confirmation, consensus under postpositivism is a discursive construction of competing views (Danziger 1995). For postpositivists, the empirical data of a neopositivist consensus is turned into knowledge only through interpretative interaction with the other perspectives. Only by examining the data through conflicting frameworks or standpoints can the hidden suppositions that give it meaning be uncovered or exposed. For the postpositivist, the crucial debate in politics is seldom over data per se, but rather the underlying assumptions that organize it. Such deliberations produce new understandings in a process better framed as a "learned conversation" than the pursuit of empirical proof. Emphasis shifts from the narrow concerns of empirical theory to the development of "a rich perspective" on human affairs (Toulmin 1990, 27).

Knowledge, in this evolving conversation, is understood more accurately as consensually "accepted belief" than as proof or demonstration.[8] Such beliefs emerge through an interpretive forging of theoretical assumptions, analytical criteria, and empirical tests warranted by scholarly communities (Laudan 1977). Instead of understanding these beliefs as the empirical outcomes of intersubjectively reliable tests, the postpositivist sees them as the product of a chain of interpretive judgments, both social and technical, arrived at by researchers in particular times and places (Bernstein 1983). From this perspective, social scientific theories can be understood as assemblages of theoretical presuppositions, empirical data, research practices, interpretive judgments, voices, and social strategies (Deleuze et al. 1987). One of the primary strengths of a theory, in this respect, is its ability to establish discursive connections and identify equivalencies among otherwise disparate elements, as well as to incorporate new components.

Although the standards of relevance and assessment of a postpositivist social science cannot be formulated as fixed methodological principles, this should not be taken to mean such research lacks rigor. In many ways, the adoption of a multimethodological approach opens the door to a more subtle and complex form of rigor. Instead of narrowly concentrating on the rules of research design, combined with statistical analysis (which usually passes for empirical rigor), the postpositivist approach brings into play a multimethodological range of intellectual skills, both qualitative and quantitative. Basic is the recognition that an epistemology that defines rationality in terms of one technique, be it logical deduction or empirical falsification, is too narrow to encompass the multiple forms of rationality manifested in scientific practices.[9] The interpretive judgments that are characteristic of every phase of scientific investigation, as well as the cumulative weighing of evidence and argument, are too rich and various to be captured by the rules governing inductive or deductive logic (Collins 1992). For this reason, postpositivism substitutes the formal logic of science with the informal deliberative framework of practical reason.

Practical Reason as Reasoning-in-Context

The search for the postpositivist alternative begins with the recognition that the formal models of deductive and inductive reason misrepresent both the scientific and practical modes of reason. As Scriven (1987, 7) argues, the classical models of inductive and deductive reason provide "inadequate and in fact seriously misleading accounts of most practical and academic reasoning." Most of such reason — for example, that of the judge, the surgeon, or the historian — has been falsely assessed as an incomplete version of the deductive reasoning of logic or mathematics, long aspired to in social scientific explanation. They are more appropriately conceptualized as forms of informal logic with their own rules and procedures. In pursuit of an alternative methodological framework, postpositivists have returned to the Aristotelian conception of "phronesis," or the informal logic of practical reason.

Informal logic, designed to probe both the incompleteness and imprecision of existing knowledge, reconceptualizes our understanding of evidence and verification in investigations that have been either neglected or mistreated by formal logics (Scriven 1987). Countering social science's emphasis on generalizations, informal logic probes the argument-as-given rather than attempting to fit or reconstruct it into the confining frameworks of deduction and induction. Toward this end, it emphasizes an assessment of the problem in its particular context, seeking to decide which approaches are most relevant to the inquiry at hand.

By expanding the scope of reasoned argumentation, the informal logic of practical reason offers a logical framework for developing a multimethodological perspective. Most fundamental to practical reason is the recognition that the kinds of arguments relevant to different issues depend on the nature of those issues: what is reasonable in clinical medicine or jurisprudence is judged in terms different from what is "logical" in geometrical theory or physics (Toulmin 1990). Basic to such judgment is a sensitivity to the contextual circumstances of an issue or problem. Practical reason, as such, distinguishes contextually between the world of theory, the mastery of techniques, and the experiential wisdom needed to put techniques to work in concrete cases. In doing so, practical reason supplies a conception of reason that more accurately corresponds to the forms of rationality exhibited in real-world policy analysis and implementation, concerns inherently centered around an effort to connect theory and techniques to concrete cases.

Practical deliberation thus seeks to bring a wider range of evidence and arguments to bear on the particular problem or position under investigation. As Hawkesworth (1988, 54) explains, "The reasons offered in support of alternatives marshall evidence, organize data, apply various criteria of explanation, address multiple levels of analysis with varying degrees of abstraction, and employ divergent strategies of argumentation." But the reasons given to support "the rejection of one theory do not constitute absolute proof of the validity of an alternative theory" (54). Through the processes of deliberation and debate, a consensus emerges among particular researchers concerning what will be taken as valid explanation. Although the choice is sustained by reasons that can be articulated and advanced as support for the inadequacy of alternative interpretations, it is the practical judgment of the community of researchers, not the data themselves, that establishes the accepted explanation. Such practical judgments, rather than supposed reliance on proof, provide the mechanism for not only identifying the incompetent charlatan but investigating the more subtle errors in our sophisticated approximations of reality. To be sure, the informal logic of practical reason cannot guarantee the eternal verity of particular conclusions, but the social rationality of the process is far from haphazard or illogical. Most important, it supplies us with a way of probing the much neglected contextual dependence of most forms of argumentation (Scriven 1987).

As a contextual mode of reason, practical reason takes place within a hermeneutic "circle of reason" (Bernstein 1983). To probe specific propositions requires that others must be held constant. Such analysis, however, always occurs within a context of reference grounded in other sets of presuppositions. Moving outside each framework to examine it from yet new frames permits the inquirer to step beyond the limits of his or her own languages and theories, experiences, and expectations. This increases the number of relevant perspectives but need not lead to a hopeless relativism, as is often thought. Because the hermeneutic process is typically initiated by external stimuli in the object-oriented world, critical interpretations are "world-guided" and can never be altogether detached from the world (Williams 1985, 140). That is, in the words of Bernstein (1983, 135), the process "is 'object' oriented in the sense that it directs us to the texts, institutions, practices, or forms of life that we are seeking to understand." Such empirical stimuli cannot compel definitive interpretations, as the empiricist would have us believe, but they do work to limit the number of plausible interpretations. While the possibility of multiple interpretations remains, there are thus boundaries or limits to what can count.

At minimum, an interpretation that bears no plausible relationship to the object world has to be rejected.

Given the limits imposed by fallibility and contingency, the informal logic of practical reason speaks directly to the kinds of questions confronted in most political and policy inquiry. Bringing together the full range of cognitive strategies employed in such inquiry, it judges both the application and results of such methods in terms of the contexts to which they are applied. Recognizing social context to be a theoretical construct, as well as the underdetermination of our available knowledge, practical deliberation probes the competing understandings of a particular problem and the range of methods appropriate to investigating them. Framing the analysis around the underlying presuppositions, postpositivist analysis seeks to anticipate and draw out the multiple interpretations that bear on the explanation of social and political propositions.

From this perspective, the postpositivist expert must function as an interpretive mediator operating between the available analytical frameworks of social science and competing local perspectives. In the process, a set of criteria is consensually derived from the confrontation of perspectives (Innes 1990). Such criteria are employed to organize a dialectical exchange that can be likened to a "conversation in which the horizons of both the social scientists and the local citizens are extended through confrontation with one another" (Dryzek 1982, 322). Thus interactions among analysts, citizens, and policy makers are restructured as a conversation with many voices (Park 1993). Given the reduced distance between the experts and the citizens, the role of both can be redefined. In effect, whereas the citizen becomes the "popular scientist," the analyst takes on the role of a "specialized citizen."

As specialized citizen, the expert can never remove political choice from the analytical process. Such analysis can, as Hawkesworth (1988, 191–92) makes clear, "expand the scope of political possibility by increasing awareness of the dimensions of contestation, and hence, the range of choice, but it cannot dictate what is to be done in a particular policy domain." The "rational judgment" of the analyst can never substitute for the choices of the political community. Postpositivism thus requires a participatory practice of democracy. "By encouraging policy-makers and citizens to engage in rational deliberation upon the options confronting the political community," as Hawkesworth (1988, 193) puts it, the postpositivist analysis "can contribute to an understanding of politics which entails collective decision-making about a determinate way of life."

Local knowledge and participatory inquiry are thus an inherent part of a postpositivist practice. We have seen that the postpositivist critique not only accounts for the normative limits of conventional practices but offers an interpretive model of practical discourse geared to the normative contexts of social action. This practical model of reason not only situates empirical research within a larger framework of normative concerns but also provides an alternative perspective on the problem of competing methodologies.

The Problem of Relativism

But do these multiple interpretations lead to a hopeless relativism? Not as conventionally maintained. For the postpositivist, the question is an outmoded relic of positivist epistemology. The strategy is simply to turn the question around and to charge the positivist with erasing the very social contexts that make meaningful judgments possible. The first step is to show how the pursuit of universal knowledge necessarily depends on the systematic narrowing and obscuring of social categories. In the name of an abstract language, it eliminates or subjugates local knowledge. The second step is to illustrate the ways in which such local knowledges are subordinated to, or substituted by, the social categories of the elites who establish the "official meanings" of the dominant society. The positivist critique thus falls into its own trap; universal knowledge is itself an ideology built on relative social concepts, the concepts of those on top.

From a postpositivist perspective, the issue of relativism can be re-defined as a question of location rather than criteria. Following Haraway (1991), the key practice that grounds all knowledge is "position," or where to see from. A way of seeing, or "vision," to use her term, involves "a politics of positioning." Rejecting the possibility of a uni versal vantage point, Haraway argues that only "the dominators at the top of the social structure can see themselves as self-identical, unmarked, disembodied, unmediated, [or] transcendent" (1991, 191). At the bottom of the social hierarchy, the political struggles of the oppressed are invariably grounded in a politics of positioning; they emphasize the capacity to see from the peripheries. To be sure, such groups have often romanticized the vision of the less powerful, failing to recognize that such positions are themselves never exempt from critical examination. But from an epistemological perspective, according to Haraway (1991, 188–201), the periphery provides the key. Peripheral positions, in her words, "are to be preferred because they promise more adequate, sustained, objective, transforming accounts" of the world. Similarly, Foucault (1972) urges us to focus on the "marginal man" standing outside the mainstream of events. For him local resistances hold key insights into the real nature of the system. Because of their partiality, subjugated vantage points can remain as vigilantly hostile to the various forms of relativism as the most explicitly totalizing claims to scientific authority. Thus the alternative to the single-visioned relativism of universal theory is the partial, locatable, critical knowledge capable of sustaining the kinds of connections that we call solidarity in politics and shared conversations in epistemology. Knowledge claims that are "unlocatable," Haraway argues, are irresponsible, as they cannot be called directly into account. From this dialectical perspective, it is precisely the politics and epistemology of partial perspectives that make possible sustained, objective inquiry. Struggles over what constitutes the rational, objective account are always struggles over how to see.

Basic to participatory expertise, as we shall see in chapter 9, is just such a shift in the way of seeing. Professional experts, it will be recalled, have been criticized for their accommodation to elite beliefs and values. Because of the experts' middle-level position in the social structure, they have too often accepted the basic premises of corporate-bureaucratic domination. The professional-client hierarchy has thus been denounced as serving — both wittingly and unwittingly — to impose systems imperatives on the intermediate and local levels of the social system. Emerging as a part of this critique, participatory expertise can most fundamentally be conceptualized as a shift in the professional's position within the structure. The participatory professional operates from the local context on its own terms, rather than prescribing premises from above. Such research works to facilitate the development of an alternative understanding based on the experiences of those in the situational context. It is an exercise in the politics of positioning.

Policy Inquiry: Empirical Analysis in Normative Context

Postpositivist policy analysis, as Hawkesworth (1988, 191) puts it, "derives its justificatory force from its capacity to illuminate the contentious dimensions of policy questions, to explain the intractability of policy debates, to demonstrate the deficiencies of alternative policy proposals, to identify the defects of supporting arguments, and to elucidate the political implications of contending prescriptions." Through a systematic examination of the contentious or problematic assumptions underlying "the constitution of perception, cognition, facticity, evidence, arguments, explanations, and options, post-positivist policy analysts can surpass positivist policy analysis because more is examined and less is assumed."[10]

These kinds of epistemological concerns are quite different from those normally encountered in policy inquiry and are not at all well received in some quarters. In most cases, the critical question rests on the status of the empirical: what happens to empirical research in such a discursive approach? Although many postpositivist writers are not clear on this question, postpositivism in no way necessitates the rejection of empirical investigation. Indeed, rather than abandoning the empirical, the approach adopted here concerns its relationship to the normative. How is the empirical situated in a larger set of normative concerns that give its findings meaning? This is the critical question that must be addressed.

More specifically, then, what does it mean to say that policy analysis should embrace this postpositivist "argumentative turn" (Fischer and Forester 1993). As we have seen, scientific conclusions are in fact arguments designed to convince other scientists to see a particular phenomenon one way or another. Although findings are traditionally put forth in the language of empirical verification — advanced as evidence that a proposition is true or false — quantitative data are only a part of a broader set of factors that go into structuring the conclusions.[11] As we have discussed, behind these conclusions are a multitude of interpretive judgments, both social and technical. The conclusion as a whole can in fact be understood better as an argument rather than as an inductive or deductive proof.

How can we conceptualize the findings or social scientific findings or conclusions of policy analysis as arguments? One of the first policy scholars to call for such a reorientation is Majone. The structure of a policy argument, Majone (1989, 63) writes, is typically a complex blend of factual statements, interpretations, opinions, and evaluations. The argument provides the links that forge the connections among these components and the conclusions of an analysis. Having recognized the epistemological shift, however, Majone neglects to account sufficiently for the normative dimensions that intervene between findings and conclusions. From the preceding discussion, we now can formulate the task as a matter of establishing interconnections among the empirical data, normative assumptions that structure our understandings of the social world, the interpretive judgments involved in the data-collection process, the particular circumstances of a situational context (in which the findings are generated or the prescriptions applied), and the specific conclusions. The scientific acceptability of the conclusions depends ultimately on the full range of interconnections, not just the empirical findings. Although neopositivists see their approach as more rigorous and therefore superior to less-empirical, less-deductive methods, the postpositivist model of policy argumentation actually makes the task more demanding and complex. Not only does it encompass the logic of empirical falsification, but it includes the equally sophisticated normative questions within which it operates. The researcher still collects the data but now has to situate or include it in the interpretive framework that gives it meaning. No longer is it possible to contend that such normative investigations can be ignored, as if they somehow relate to another field of inquiry.

Elsewhere I have suggested a multimethodological framework for integrating these concerns. In *Evaluating Public Policy*, I have offered a logic of four interrelated discourses that outline the concerns of a more comprehensively rational policy evaluation. Extending from the concrete questions concerning the efficiency of a program up through its situational context and the societal system to the abstract normative questions concerning the impact of a policy on a particular way of life, the scheme illustrates how empirical concerns can be brought to bear on the full range of normative questions.

Conclusion

This chapter has examined the postpositivist foundations of a more democratic or participatory approach to professional inquiry. After having outlined the failures of neopositivism, we saw how the postpositivist critique not only accounts for these failures but also offered an alternative approach. The alternative is an interpretive model of practical discourse geared to the normative contexts of social action. From this perspective, knowledge cannot be understood in terms of abstract, "objective" empirical theories and their statistical variables; it is seen to be anchored in interpretive social understandings. Situationally grounded in particular contexts, knowledge creation thus always involves a social negotiation. As such, knowledge need not be — and often cannot be — generalizable beyond the particular context to which it pertains.

The practical model of reason not only situates empirical research within a larger framework of normative concerns but also offers an alternative perspective on the problem of epistemological relativism. By stressing the situational context of social inquiry, practical reason speaks directly to important limitations of the empirical social sciences and offers an epistemological framework for the practice of participatory inquiry, to which we turn in part 3. Rather than a step backward, as empiricists would have us believe, postpositivism's emphases on practical reason and local knowledge are essential components of the step toward a more socially relevant mode of inquiry.

Environmental Politics in the Public Sphere

Technical versus Cultural Rationality

Science is near to being *the* source of cognitive authority: anyone who would be widely believed and trusted as an interpreter of nature needs a license from the scientific community.
— Barry Barnes and David Edge

The three chapters in part II focus on the role of science in environmental policy making, its problems as a technocratic decision strategy, and the political response of the environmental movement, in particular the movement's emphasis on sociocultural rather than technical reason. In doing so, these chapters examine more concretely the limits of the conventional neopositivist conception of science in environmental decision making and illustrate how the sociocultural dimension central to the postpositivism has been brought into play by the environmental movement.

Chapter 5 offers a portrait of the double-edged role of science and technology regarding the environment, serving as both a cause of degradation and a source of its identification. In the process, the discussion examines the conflicts between science and politics to which this tension has given rise in environmental struggles. Basic to this problem has been that the dominant technocratic orientation in environmental policy making — reflected in particular in the methods of risk assessment and risk-benefit analysis — has been fraught with empirical uncertainties that have led to a politics of expertise and counterexpertise among industry and environmental scientists. The chapter discusses the ways this indeterminacy has played itself out in both environmental bureaucracies and the courts, including the search for institutional solutions such as science courts, risk communication, and environmental mediation.

Chapter 6 explores the environmental movement's ambivalent-to-hostile relationship to science and technology and examines in more detail the movement's political responses to the technocratic orientation of the dominant environmental institutions. For many in the movement, the solution to environmental problems is to be found in a less-technocratic, if not an altogether different, kind of science coupled to greater citizen involvement. Emphasizing participatory decision making, the progressive wings of the movement have seen environmental democracy as a strategy to revitalize civic society and its normative core, the public sphere. Here the call has been for a "metapolitics" that would replace conventional interest group politics with an emphasis on the sociocultural values and consumer lifestyles underlying the environmental crisis. In this view, the environmental problem is more a crisis of our social institutions and cultural practices than a material question of too much pollution. This new "cultural politics," advanced by groups such as the environmental justice wing of the movement, holds out a radical alternative to scientific decision making.

Chapter 7 delves into the cultural politics of the environmental movement and its postpositivist critique of technocratic risk assessment. Toward this end, the concept of "cultural rationality" is introduced and explored through the politics of NIMBY (not in my backyard), a community response to anxiety and distrust about toxic hazards. Cultural rationality and its emphasis on case-specific social processes is not only fundamental to citizen decision making but also the essence of rationality in certain types of decisions, especially those involving scientific and social uncertainty. Such rationality, in this view, has to become an integral part of risk assessment. Thus the challenge ahead is not just *more* science but rather how to better understand the *interactions* between science and ideological belief systems — technical facts and cultural values — and most importantly how to systematically integrate them in a more comprehensive analysis. Indeed, as the discussion demonstrates, citizens bring a missing element to the risk inquiry process.

5. Science and Politics in Environmental Regulation: *The Politicization of Expertise*

For all [the] effort and all its presumed usefulness, I cannot
identify a single social science finding or idea that is undeniably
indispensable in any social task or effort. — Charles Lindblom

Tensions between science and politics have been intrinsic to environmental struggles from the outset. On the one hand, science and technology have been identified closely with the major causes of environmental degradation; on the other, they have served as the primary methods for both detecting environmental problems and searching for effective solutions. This has given rise to a more or less technocratic orientation to policy making that is today characteristic of environmental regulation. At the same time, the environmental movement has long adopted a critical stance toward science and technology. For many environmentalists, the emergence of a technocratic form of regulation is as much the problem as the technologies that create the pollution in the first place. This has often led the movement to call for more democratic forms of science and technology policy making as the basis for a sustainable society. As seen in the last chapter, it has at times become quite fashionable to speak of an alternative science as part of a genuine "ecological democracy." This chapter examines the role of environmental science, in particular its politicization, before turning in the next chapter to the question of public participation.

Science, Technology, and the Environment

Science and technology are fundamental to environmental politics and policy making. While environmental policy making shares common characteristics with other social problems, there is a notable difference. Whereas social problems typically draw much of their rhetorical power from moral discourse (e.g., Should women get the same pay as men? Should the homeless sleep in the park?), environmental problems turn much more on arguments about "facts." Problems such as global warming, while morally charged, tend to be more directly tied to scientific findings and claims. Although they are generally traceable to human agents, environmental problems have an imposing physicality compared to other social problems, which are more often rooted in social and personal concerns that are converted into public issues (Harrigan 1995).

To understand the political dynamics of science in the environmental regulatory process, we need to take into account its double-edged role. On one level, the relationship is fairly direct and obvious. Most environmental problems are—in one way or another—the negative consequences of techno-industrial society. Chemistry, for example, is at the root of the toxic waste problem, one of the most severe environmental challenges of our time. Automobile engineering is basic to air pollution and the greenhouse effect. Physics underlies the dangers of nuclear radiation, and so on. On another level, however, the environmental crisis is in significant part a scientific discovery. Since the beginning of modern day environmentalism, science has in most ways been responsible for detecting and measuring the very problems to which scientific technologies have given rise. Science's ability to better detect cancers resulting from toxic wastes has led to a more refined understanding of the relationship between hazardous chemicals and illness. Climatic observations have discovered the otherwise invisible ozone hole and the dangers of global warming. For many, further reliance on science and technology offers the only viable strategy for finding effective solutions to environmental problems. Whether or not it holds the key, given the technical dimensions of environmental problems, science will clearly continue to play a central role. The critical question concerns the nature of the role. Whether in technocratic form, or in some alternative mode, science and technology will have an important role in dealing with the environmental crisis.

Technocratic Environmentalism

To adequately appreciate this connection between environment, science, and technology, one need only observe the science-based nature of environmental policy making. Indeed, environmental policy making has given rise to a "new model of scientific regulation" (Schmandt 1984). Scientific and technological determinations have become the primary standards by which substantive regulatory decisions affecting environmental quality are reached. This dependence on scientific criteria in environmental regulatory rule making can be witnessed in countless ways. Since 1970 most federal environmental programs have required the implementing agencies to specify empirical standards for ambient air and water quality, to identify the appropriate control technologies for pollution sources, to define acceptable risks from exposure to toxic hazards or substances, and to make a multitude of other environmentally relevant technical judgments. In effect, federal statutes thrust on governmental agencies the burden of scientifically defining acceptable levels of pollution, pollution abatement, and risk of exposure to environmental pollutants. For this purpose, federal and state environmental protection agencies have developed extensive mechanisms for generating scientific information, from internal expertise to external advisory boards (Jasanoff 1990).

Accompanying this development has been a firm belief that good science can show us the way. This emphasis — if not faith — in scientific analysis was most clearly expressed by William Ruckelshaus (1991, 54), former director of the Environmental Protection Agency (EPA), when he wrote:

> We are now in a troubled and emotional period for pollution control; many communities are gripped by something approaching panic, and the public discussion is dominated by personalities rather than substance. . . . I believe that part of the solution to our distress lies with the idea that disciplined minds can grapple with ignorance and sometimes win: the idea of science. We will not recover our equilibrium without

a concerted effort to more effectively engage the scientific community. . . . I need the help of scientists.

Nothing better illustrates the reliance on science than the EPA's use of technocratic decision techniques. In the 1980s, the agency established the technique of "risk-benefit analysis" as the basic tool for agency decision making. All agencies are required to rigorously calculate the monetarized costs of achieving a particular level of risk protection against the expected dollar benefits associated with each proposed regulation (see appendix A). Each new proposal must pass such a risk-benefit analysis test before it can be considered and adopted as law. As formally introduced—if not always practiced—the decisive test of any new environmental regulation is the ability of risk-benefit analysts to show that it efficiently offers a net benefit to society. Underlying this utilitarian calculus is the idea that empirically measured consequences rather than social preferences provide the appropriate standards for policy making. Grounded in a post-Enlightenment faith in science and technology, the technocratic use of risk-benefit analysis is built on the conviction that science provides the foundation of rational decision making. But this faith in scientific analysis proved to be much more complicated than people like Ruckelshaus recognized. In fact, science has very often only intensified the very politics that those who turned to it sought to circumvent. How could that happen?

The Politicization of Science

The double-edged role of science in environmental policy making has caught scientific decision making in a compromised, if not contradictory, situation that has given rise to sharp political conflicts between citizens and experts. On the one hand, scientific technologies have emerged as the modern version of the Faustian bargain. Nuclear power, for example, holds out the possibility of an abundant source of energy, but at the same time it threatens the future of modern civilization. Citizens' environmental movements have organized themselves around protection from the risks presented by such technologies. On the other hand, the struggle against such risks has typically lifted the technical questions concerning the actual risks to the center of environmental debate. And here citizens have found themselves disadvantaged. Often they have discovered themselves hindered from participating in environmental deliberations. Given that it was an active citizenry that brought about the contemporary environmental movement in the first place, this political dynamic is of no minor consequence.

Science clearly contributed to the early phases of environmentalism — for example, the discovery of the nonviability of bird eggs exposed to pesticides or the presence of radioactive strontium in reindeer meat. But science can scarcely be credited with having initiated the environmental movement. This movement resulted from citizen concerns. Indeed, many of the initial environmental problems could simply be identified by the eye or nose of the layperson. In this period of the movement, in fact, scientific technologies were often demonized. Technologies such as the automobile and nuclear power were denounced as sources of environmental degradation, giving rise to a strong anti-science and -technology movement that today still remains a part of environmental movement in general. Emphasizing the philosophy of "small is beautiful," a highly vocal segment of the movement calls for a return to a simpler life based on "appropriate technologies."

In the second phase of environmentalism, however, scientists themselves took center stage, including "movement scientists." In these years, new problems were often beyond immediate sensory detection. The knowledge of more sophisticated levels of physical and biological complexity that now characterizes environmental issues evolved with each new phase of environmentalism. Uncovering the infinitesimal and invisible sources of danger and risk required the intricate measurement instruments of science. For example, it took extensive scientific research and debate to figure out what causes acid rain. What is more, many of these newer problems pose even greater threats to human health and well-being than earlier concerns about polluted air and water. Now concerns such as the ozone hole and global warming threaten the very future of life on the planet as we know it. The result has been an increasingly technocratic environmentalism, in the environmental movement as well as the corridors of governmental decision making.

In many ways, this increasingly technocratic environmentalism has been a product of the very success of the early movement. Whereas the environmentalism of the 1960s was a kind of street politics, the environmental politics of the middle 1970s and 1980s spoke more through the technical languages of environmental management. Once established on the political agenda, the struggle over environmental policy shifted from the public arenas of protest to the institutional arenas of expertise, in particular to the governmental administrative arenas. Here the focus turned to difficult technical questions related to goals and solutions. In the process, environmental decision making became increasingly embedded in the technocratic languages of environmental impact assessment, cost-benefit analysis, technology assessment, and risk-benefit analysis. Environmental politics, in the process, began to be played out as encounters among industry-oriented experts and environmental counterexperts arguing among themselves over the merits of competing assessments.

Scientific Indeterminacy

The Politics of Counterexpertise

Political and economic leaders turned to science in search of a way out of complicated environmental conflicts. Science was to serve as a firm basis for making and justifying reliable decisions. But this strategy proved to be much less promising for environmental policy than expected (Barker and Peters 1993). For science itself, it turned at times into a kind of nightmare. What its proponents didn't realize was how "underdetermined" science would prove to be, to use the phrase of Collingsridge and Reeves (1986). Whereas those who turned to science and scientific decision-making methods assumed an "overdetermined" model—namely, that science could in fact answer the questions in such a way that would eliminate or at least significantly reduce political conflict among affected parties—the actual experience was quite different. Environmental politics was to entail a series of confrontations between science and society that raised questions about science itself, questions over both the direction of scientific work and the assessment of its results.

The basic problem was straightforward. In its application to policy decisions, science discovered that it could not answer the environmental questions with enough precision to be decisive. Indeed, it often tended to raise more new questions than it could answer. And it was just this uncertainty that opened up — unintentionally — the space for a politicization of science. In short, the outcome was the opposite of the anticipated benefits.

This politicization of science was the direct result of its indeterminancy. Environmental issues pushed science into realms where the evidence was either limited or mixed. This indeterminant nature of the relevant scientific questions led to strong scientific and political disagreements. First, the fact that scientists could not settle such complex questions opened the door to competing interpretations of the same phenomena. Disputes over the health effects of radiation, dioxin, and lead revealed that something besides "objectivity" had come into play. This was especially the case in terms of the long-term, chronic effects of low-level exposures that required more sensitive measurements than had yet been perfected. The myriad complex relationships entailed in the assessment of such risks were often subject to competing interpretations. At the same time, whereas knowledge was extensive about some subjects — for example, cancer — there was less about others, such as sperm defects. Knowledge might be relatively extensive about some people (such as healthy adult male workers), but less so about others (such as younger and older people). In short, scientific work had expanded the realm of what was *unknown* far more rapidly than it had the realm of the known.

Scientific conflicts over environmental issues have often also resulted from the diverging perspectives of the different disciplines. As professional specialization increasingly led to more fragmented scientific disciplines, each group of specialists came to know more and more about less and less. Hence each specialization featured its own distinctive outlook, giving rise to different types of "realities." Training and experience in a particular branch of science took on the character of personal involvement in a piece of reality to which one became emotionally attached, often making it difficult to grant other segments of reality a similar degree of importance. As demonstrated by the experiences of the EPA's Science Advisory Board, it became clear that science could mean many different things (Jasanoff 1990). Although scientists have typically argued that science's core methods are basically constant from field to field, and that scientists are thus capable of assessing research in other fields, the disputes between the different specializations revealed more fundamental disagreements. Hays (1987, 343) put it this way:

> The intensity of training and practice in one specialty created an intellectual and personal commitment to what one knew best and about which one could speak most authoritatively. Professional standing depended on the ability to describe a particular piece of reality in such a way as to convince others that one was right, and their confidence only reinforced one's emotional commitment to that reality. Specialization thus created limited, not universal, perceptions, mind-sets that gave rise to fundamental differences in the way the world of scientific reality was understood and interpreted.

All of this became especially problematic in the realm of public policy. Policy decision makers seek answers. They operate in a world of imposing time constraints, where facts can seldom be separated from values, where it can be difficult to differentiate independent and dependent variables. It is a world in which discussion about issues can seldom be kept private; it is a sphere in which people use information to suit their own purposes. Nothing, in this respect, could be more alien to the environment of the laboratory and the deliberative processes of peer review. Advancing through gradual exchange among scientists in journals and scientific meetings, academic science occasionally gives rise to sharp disputes, but they are almost always carried out without pressures to agree. Indeed, in its idealistic form, science proceeds through disagreement. The skeptical orientation defines its basic stance. But in the realm of environmental science, this relatively relaxed system began to change drastically in the 1960s. The entrance of environmental science into the world of law and public policy presented circumstances difficult for environmental science to accommodate. It was one thing for scientists to be willing to express views in the more relatively protected collegial setting of a scholarly journal or a professional meeting. It was quite another to be subjected to relentless cross-examination on a witness stand. The result has been frequent conflicts with decision makers in search of answers. For decision makers, scientific uncertainty has often proved to be a troublesome problem.

The nature of this indeterminancy gave rise to a model of regulatory policy making that Jasanoff (1990, 49–57; 1995) has labeled the "science policy paradigm." Given the inability of the various environmental science advisory boards to establish conclusive findings on which policy making could be based, a model for decision making emerged that brought together a mix of scientific and administrative-legal considerations. Evolving through congressional legislation and agency practices, the paradigm evolved around three interrelated elements, each of which deeply influenced both agency procedures for evaluating science and the structure of scientific advisory processes. The first is the notion that agencies should be permitted to make regulatory decisions on the basis of imperfect knowledge (that is, suggestive rather than conclusive evidence). The second element, a corollary of the first, is that science policy determination may be regarded as valid even if the scientific community does not universally accept it as such. And third, when experts disagree about the validity or interpretation of relevant data, the administrative agency should have the authority to resolve the dispute consistently with its legal mandate.

The model evolved in response to an unavoidable circumstance: decisions have to be made. When science alone is incapable of providing decisive answers to questions of risk, the choice among conflicting assessments still must be made. In light of this fact, according to the model, the decision should be made by the politically accountable agency in accordance with its lawful regulatory mission. Given their unique combination of technical and policy skills, the new environmental administrative agencies appeared, in principle, well qualified to make science policy determinations.

The full implications of agency authority in this area began to emerge only gradually as the courts advanced new interpretations of statutory language aimed at health, safety, and environmental protection. The science policy paradigm assumed, in theory, that high administrative discretion on the part of agencies in resolving scientific disputes would be coupled with equally high judicial deference to the agency's expert judgments. But as the courts began to exercise their obligation to monitor the agencies' substantive evidence and rationales for regulatory decisions, the theory of deference gave way to a widespread practice of judicial review of technical decisions (Jasanoff 1995). As the number of science-based regulatory decisions rose in the 1970s, the result became extremely problematic for the science policy paradigm.

Following the requirements of administrative procedures statutes, the courts can overrule decisions judged to be either insufficiently supported by the given evidence or an arbitrary or capricious abuse of discretion. In view of this mandate, the courts, especially in the early years of environmental decision making, began to take a "hard look" at the technical evidence before them. Although courts expressed no desire to make scientific findings on their own, they insisted on holding agencies accountable for a full and reasonable explanation of their technical determinations. To ensure that an agency acted "reasonably," the reviewing judge was entitled by this standard to probe into the administrators' scientific thought processes. As Jasanoff (1990, 55) explains:

> The "hard look" metaphor in this way subtly became a rationale for letting the courts themselves look hard at the scientific arguments underlying the agency decisions; in other words, it provided a vehicle for remarkably intrusive review in just those areas where deference was in principle most warranted. In a growing body of cases, particularly those involving decisions not to regulate, the courts set aside decisions because they found the agency's scientific record or reasoning inadequate.

To simplify a complicated story, the consequence of the court's behavior was to make public the general inadequacy of the available knowledge. In particular, it revealed the number of interpretive decisions involved in scientific practices under the conditions of uncertainty (Jasanoff 1995). For the EPA, this was especially problematic, as the agency was specifically given the task of making decisions in the face of scientific uncertainty (Cranor 1993). Indeed, the science policy paradigm was established to compensate for this fact. Agency officials were to collect the best expert judgments and then assemble them into reasonable rules that fit within the parameters of the agency's legal mandate. But now the courts were criticizing the agency for playing the role it had been assigned. For science and the advisory committees, this situation brought to public attention and scrutiny their limited ability to serve this decision-making function.

These court opinions opened a large hole through which the opponents of environmental regulation could take up the process of counterexpertise.[1] Because little could be "proven" in the conventional understanding of the term, all one had to do was find fault with a particular dimension of a study, a strategy most easily pursued through criticisms of experimental and statistical methods. Later the courts retreated from this position and permitted the agency to balance technical findings with more interpretive social and political considerations, as specified by Congress. By that time, however, the cat was out of the bag. As Jasanoff (1990, 59) writes, "Inconsistent decisions and wavering judicial support underscored the political fragility of the science policy paradigm and added weight to industry's demands for better quality controls on regulatory science."

The chemical industry, for example, had never conceded that agencies should have as much discretion in making risk assessment analyses as the standard understanding of the science paradigm offered them. Indeed, controversies over substances such as benzene and formaldehyde represented, at bottom, a tug of war among regulators, chemical producers, and public interest groups about the extent to which the paradigm should be used to resolve disagreements over scientific uncertainty. Industrial opposition to particular regulatory rules rapidly transformed into demands for a return to more technocratic processes for deciding regulatory questions. Convinced that a techno-analytic approach would lead to more scientifically conservative decisions — and hence decisions more sympathetic to business — opponents of the paradigm set out scientific risk assessment guidelines, suitable for resolution by accredited expert bodies such as the National Academy of Sciences. For industry, emphasis on agency discretion was to be reduced and placed on the process of scientific peer review.

Although this message — "Leave science to the scientists" — was superficially appealing, it failed to address the underlying problem of defining what counts as "science" in areas of methodological uncertainty and political conflict. The combination of indeterminancy and political pressure gave rise to the politics of expertise and counterexpertise (or political "antidotes," in Beck's words) that became rampant in environmental politics, employed by both industry and the environmental movement. In a political world, this meant that those who didn't like the outcomes for reasons of interest could easily find or employ different experts to present the other side of the issue. If one wished to discredit a given study, it was easy to question its experimental or statistical methodology. Regardless of the merits of the study, one could argue with little difficulty that it failed to take into account other possible causes of a given effect. The intervening or confounding variable was of special interest to those who sought to draw attention away from a potential cause of contamination and direct it toward another. As Jasanoff (1990, 59–60) observes, "The struggle for control over regulatory policy was thus played out in part on the fields of discourse, as terms like 'science,' 'policy' and even 'peer review' were redefined to fit different conceptions of the relationship between science and power."

Given the holes in the existing data, environmental protagonists could either demand higher levels of proof or draw conclusions on the basis of lower levels of certainty. Industry leaders pressed this process of counterexpertise quite forcefully. Whereas environmentally oriented advisers argued that there was enough evidence to justify a conclusion, given the seriousness of the potential consequences, industry supported experts who argued that there was not yet enough solid evidence to justify the economic costs of a regulatory rule. Industry, in short, brought to legislative, administrative, and court hearings scientists who insisted on high levels of proof of harm. They criticized environmentalists' margins of safety as resting on unreasonable notions about plausible but not yet fully discovered unknowns, thus requiring unjustified degrees of protection.[2] Indeed, it was this industry argument that took the driver's seat in the 1980s with the arrival of the Reagan administration. Taking their lead from industry, this administration put strong pressure on the EPA to adjust its public regulatory decision-making standards accordingly. Via executive order, the Reagan administration ordered that all regulatory decisions be judged by a cost-benefit analysis, which in the case of environmental regulations got translated into risk-benefit analysis.

Such disputes, moreover, were often played out in intensely emotional terms. "Conventional" scientists, on the one hand, argued that there was "absolutely no proof" that a given pollutant was harmful and derided dissenting scientists as tainted by nonscientific and emotional tendencies. "Frontier" scientists, on the other hand, argued that those who demanded high levels of proof had their own unscientific commitments, whether in their predispositions or in their loyalties to the industries for which they worked or from which they received financial support.

Although the process was politically complex, it is clear that the most important determinants of this politics were money and interests (Hays 1987). Without great difficulty, one could trace respective positions to those who funded the research. The identification of problems for research, the selection of scientists to conduct it, and the construction of research designs all too often stemmed from choices made by those in private industry and government who supplied the research funds. Scientific research and advice were thus frequently shaped or influenced rather directly by the financial stake scientists had in its acceptance. In recognition of this relationship, major advisory boards took steps to exclude from their review committees scientists who were stockholders in either the companies for which they worked or those that financed activities the committee was asked to review. Later this exclusion was extended to include a scientist's role as employee of a company or agency with a stake in the decision.

The relationship in sponsored research was subtler. For some scientists, the main questions raised by sponsorship involved their right as independent investigators to publish freely the results of their research. Many private corporations would not permit this; they often requested scientists to formally agree to refrain from such publications. In several cases, negotiations between industry sponsors and scientists were broken off because such agreements were publicized among scientists as warnings of what could happen. Both scientists and corporations sought to mute this relationship by "laundering funds" through research institutes, but many observers found that such arrangements neither modified corporate expectations nor changed the understandings of these implications by scientists.

Essential for environmental politics was an expansion of funding sources. As researchers were able to obtain more and more monies from other funding sources — public health and other governmental agencies in particular — industry was increasingly less able to control the flow of data. New researchers, obtaining funding from public health agencies, helped to broaden the spectrum of views beyond those of industry. According to Hays (1987), major policy changes can be attributed to these shifting patterns of information flows. Such shifts were basic to the advance of the environmental movement and the development of environmental counterexpertise.

The issue of bias also directed considerable attention to the appointment processes of various technical advisory committees. As it became clear that the basic factor in determining the outcomes of the advisory process rested on the composition of the membership of scientific advisory groups, each group labored to get its own like-minded scientists on the committees (Jasanoff 1990). This struggle often emerged along party lines, with Republicans advocating industry-oriented scientists and Democrats putting forward scientists acceptable to the environmental movement.

But the politics of the environmental decision-making process scarcely stopped there. The technical advisory groups in the environmental protection agencies that wrote the regulations seldom relied on the scientific evidence alone. Such committees comprised a mix of environmental scientists, economists, and lawyers. Although these committees largely started out as technical decision-making groups, economists and lawyers were added as the criticisms of environmental regulations increasingly centered on economic factors and court decisions. Regardless of the technical merits of a proposed regulation, it became necessary to have lawyers present, as each decision was almost always challenged in the courts. Indeed, new regulations are typically petitioned in court on the very day, if not hour, of their release. And it was in the argumentation of the lawyers that these interpretive gaps in science were trotted out for public display. This came to mean that there was little point in issuing a regulation if it could not stand the juridical tests that it would confront in court.

The consequences of this politicization of expertise were at times quite devastating for the scientific community. Insofar as the public witnessed these debates, they came to see the subjective and political sides of science. That different experts could interpret the same phenomena differently is not so problematic for science itself. But in environmental policy making it seriously tarnished the technocratic "overdetermined" understanding of science. Beyond this realization, other political dimensions of the process emerged as well. Given the opportunity to watch countless replays of the same policy debates, citizens could start to recognize that almost always the scientists employed by industry and the environmental movement respectively took the same adversarial positions. It was clearly one thing to say that scientists could interpret complex data differently, but quite something else to recognize the *patterns* of interpretation that followed specific political lines. Stated differently, why did the scientists employed by industry always argue that there was not yet enough information to say that something had been proven? Why did the environmentalists always argue that the results were clear enough to take action?

It didn't take environmentalists long in this process of counterexpertise to discover that many — maybe even most — professionals tended to be biased against their environmental positions. The lion's share of the interpretive judgments made by professionals has tended to work in favor of industry, even if indirectly. Environmentalists, for this reason, have had much more trouble finding the support of the professional communities. It is, of course, risky to characterize an entire community. But it is fair to say that scientists and professionals as a whole have served as a relatively conservative force in the environmental politics that emerged after the 1960s. Although professional experts in environmentally related fields have generally taken environmental values to be legitimate social objectives, they have not as a group been so friendly to the environmental movement per se. While they have not necessarily aligned themselves to business opposition, they have largely assumed a professional stance of caution that has done little to counter this opposition. They have as a group tended to believe that environmental advocates were pushing forward their objectives further and faster than the nation's resources would permit. Seldom have scientists called for a rapid expansion of the frontiers of environmental policy or research. More often than not, they have viewed the movement as pushing the issue to excess, expressing the problem in emotional rather than rational terms. Even though they have tended to advance these arguments in terms of the public interest, such arguments worked in effect to support business groups who opposed environmental action for more self-interested reasons. Indeed, they have often been used by industry as direct support for their hostility to environmental values.

Scientists have also been quite critical toward public intervention in their affairs. Public involvement in the scientific aspects of environmental affairs, they argue, interferes with matters best left to experts. Scientists tend to draw a distinction between informed and rational science, on the one hand, and the far less informed and often emotional public on the other. In this respect, scientists have stood behind the customary understanding of science's role in public affairs—that of disinterested and objective investigators gathering knowledge for the public benefit. Because public issues were often technical, scientists were called on to determine the facts on the grounds that they were disinterested parties to the particular outcome of inquiry and hence would be above controversy. Because science is "objective," it is seen to hold out the possibility of unifying contending sectors of society and politics, following the technocratic formula.

Basic to this view is a fundamental epistemological principle, namely, value neutrality. In professional field after field, experts argued that their role in environmental issues was—and should be—one of neutrality rather than advocacy. Indeed, if environmental experts had anything to offer, so they argued, it was their ability to substitute concrete facts for an uninformed emotional response to the problems at hand. But as we have seen here, professional expertise has scarcely played this role. Under the guise of scientific neutrality, policy experts have more often than not taken sides.

Indeed, environmental disputes have revealed that scientists make many social choices that transcend the standard explanation of what constitutes scientific objectivity. This has raised not only concerns about the relation of knowledge to interests but also questions about the inner workings of science itself. As we saw in chapter 4, these include the selection of the appropriate topics for research, the descriptive categories and criteria to apply to the evidence, the skills appropriate to the conduct of inquiry, and the assessment of diverse findings. Why have some scientists chosen one way and others another? As a result, scientists have sometimes been seen as another interested group expressing particular values and views as a result of their training, experience, and personal predilections. Simply to raise such questions was to threaten the image of science as the instrument of universal truth. Often advanced by distinguished historians and sociologists of science, such criticisms frequently had considerable impact on both scientific self-conceptions and public views about science. These questions and criticisms did considerable damage to science's privileged position as the arbiter of rationality.

To make the matter even more problematic, all of these failures became grist for the mill of a new critical postpositivist sociology of science that took root at about the same time as the environmental issue itself. As we saw in chapter 4, this new constructivist approach began systematically to explore the social dimensions built into the scientific processes itself. With origins in the much renowned work of Thomas Kuhn (1970), which showed that science progressed more through "paradigm shifts" than straight-line accumulations of evidence, this new research came to describe science as a social practice rather than a purely objective process of knowledge accumulation. These social dimensions concerned not only the broad social and political dimensions that were associated with science and technology policy generally but also the internal practices of the laboratory (Latour and Woolgar 1979). As a result, what we have generally taken as "knowledge" came to be seen in quite a different light. Rather than the product of objectives measures per se, it emerged as the outcome of consensus. In short, science could itself be construed in significant part as an interpretive activity.[3]

The Search for Alternatives

Not surprisingly, many scientists have expressed dissatisfaction with the degree to which differences within the scientific world have become public disputes. It is a development that has tarnished the public image of science and threatened its privileged status as source of rational knowledge. Many in the scientific professions and their associations have tended to understand or construe these debates as controversies over the relative competencies of different scientists and the validity of their research findings, rather than issues about the nature of science itself. Toward this end, they have searched for methods by which scientific controversies could be resolved by scientists within the confines of their own institutions. One was to strengthen the peer review process; another was to call for more careful control of appointments to advisory committees. They also suggested more formal devices such as the establishment of science courts, the development of risk communication, and the introduction of environmental mediation.

One of the most widely discussed of such proposals was that of the "science court." Instead of attempting to resolve scientific disputes associated with regulatory activities in courts or administrative law proceedings presided over by judges possessing little competence in technical matters, such decisions should be turned over to a court of scientists. Serving as "science judges," the scientists would hear the evidence of the parties to a dispute under the strict rules of procedure and cross-examination, much like those governing regular judicial proceedings, and then resolve the differences on the evidential merits of the technical arguments. In the late 1970s, an attempt was made to test this proposal experimentally but was never implemented. Too few participants believed that truly impartial scientists could be found to serve as judges (Jasanoff 1990).

The practice of risk communication was developed to better inform the public of the scientific findings related to environmental risks (National Research Council 1989). Basically, the practice was designed to address the belief that the public responded irrationally or unintelligently to scientific evidence; citizens simply didn't understand such problems well enough to make intelligent decisions. Risk communication, in this way, approached risk questions as matters for experts to decide and, as such, sought to reinforce science's privileged position as source of rational knowledge. The development of this method gave rise to a host of psychological studies focused on how people think about such matters, but most of the evidence was of little help to decision makers. Indeed, such research began to suggest that people are irrational per se but rather only followed another type of rationality, an issue to which we turn in chapter 7.

A third approach to controlling public discussion of technical issues came in the form of environmental mediation (Susskind and Ozawa 1985). Because many environmental professionals felt uncomfortable in their association with popular movements and organizations, they sought to transfer disputes to more "neutral arenas" of environmental mediation (or "alternative dispute resolution") that limited popular influence. Toward this end, the practice of environmental mediation was developed to bring together the parties to a conflict—administrative agencies and the interested and affected parties — in a setting somewhat similar to the informal judicial procedures employed in labor-management disputes. Mediation, as such, is an attempt to substantially depoliticize public environmental disagreements by enabling professional mediators to shape the context of decision making. By focusing attention on "hard facts," rather than on the social and political dimensions seen to confuse the public, mediators seek to fashion institutional decision processes in ways that contain the political "excesses" of environmentalism (K. Lee 1993).

Thus the formal goal of environmental mediation, including its application as "regulatory negotiation," is to shape consensus among disputants through a professionally facilitated process of deliberation (Fiorino 1995). Insofar as progress in resolving disputes is hampered by excessive conflict and controversy, mediation is introduced to reduce or head off incipient disputes before they emerge into hardened or irreversible political positions. In effect, this emphasis on nonpolitical processes has meant deliberative formats more favorable to scientific and technical professionals than to political activists. To facilitate this, the business community and various foundations have given substantial sums for the advance of environmental mediation (see appendix B, "Alternative Dispute Resolution").

But environmental mediation has not proven to be the magic solution it was envisioned to be. Although its proponents see the process as pursuing nonpolitical objectives, its opponents have seen it as a strategy for biasing the deliberative process in technical rather than democratic directions. Many environmentalists have come to see the practice as biased against the normative considerations underlying the environmental crisis. It didn't take long, in this respect, for the environmental movement to recognize that the subtle ideological foundations of mediation conveyed criticism of the citizen environmental movement. The movement was quick to point out that the atmosphere around environmental mediations was one of criticism of citizen action rather than of the business community. In numerous cases, the environmental movement protested such processes by walking away from the negotiation table (van den Daele 1995). Consequently, few clear-cut breakthroughs occurred as a result of the application of this method (Menkel-Meadow 1995; J. Forester 1992).

Despite the efforts of the scientific establishment to bring this situation under control, the problem remained the scientists themselves. They were the ones who disagreed, and some even joined the environmental movement. Indeed, this was one of the most important events in the development of the movement—that is, the signing on of "movement scientists," to which we turn in the next chapter.

Conclusion

This chapter has examined the tensions between science and environmental politics and their implications for the politics of risk. Science, as we saw, has been associated with both the causes of environmental degradation and the search for solutions to environmental protection. In the process, science has given rise to both a form of technocratic environmentalism and, at the same time, increasing public anxieties about science and technology. Out of this tension emerge the political outlines of Beck's risk society.

Furthermore, we have explored the ways in which the limits of scientific policy making have exacerbated these tensions. Owing to the indeterminacies of scientific analysis in complex questions of risk assessment, technocratic decision making has given rise to the problematic politics of expertise in which each side of an issue can use its own scientists to dispute the claims of the other side. Indeed, this process of counterexpertise became a regular feature in court struggles over agency regulations. Consequently, the politics of counterexpertise has led to various efforts to innovate alternative dispute procedures, including science courts, risk communication, and environmental mediation. None of these, however, have solved the problem. In significant part, the failure lies in the neopositivist understanding of science on which these innovations continue to rest. The alternative, as we have seen, is found in a turn to a constructivist understanding of science and a less technocratic, more participatory approach to decision making and consensus building, issues to which we turn in parts III and IV.

6. Confronting Experts in the Public Sphere

The Environmental Movement as Cultural Politics

[New social movements] . . . are an opposition to the effects
of power which are linked with knowledge, competence, and
qualification: struggles against the privileges of knowledge. But they
are also an opposition against secrecy, deformation and mystifying
representations imposed on people. . . . What is questioned is
the way in which knowledge circulates and functions,
its relations to power.
— Michel Foucault

Social movements, as we saw in chapter 3, are the carriers of a new
political awareness and reflection, or "reflexivity." In the case of the
environmental movement generally, as Beck argues, it is the very *social explosiveness* of hazards that has provided the critical potential
for a new political awareness about the societies in which we live. By
offering new interpretations of the threats and dangers generated by
the processes of industrialization and modernization, the environmental movement(s) seek to usher in a new "reflexive modernity" to
replace an increasingly risky, outmoded industrial society.

The Environmental Movement

Science, Participation, and Politics

Although science has been basic to the identification of the nature and size of environmental problems, science itself can scarcely take credit for their elevation to a high-level concern on the modern political agenda. This has been the role of the environmental movement, which has focused on the political marshaling of scientific findings. It has been the movement — or more precisely the various movements within environmentalism — that has brought the relevant scientific evidence and its social and economic implications to widespread public awareness.[1]

In effect, for the environmental movement, scientific findings have served in large part as ammunition for its organizational activities. The main job of the movement, as with any movement, is to organize people to get involved. This has required eliciting citizen participation, and the means of this elicitation is politics. In the course of these activities, scientific findings have been used when they have served to support or further the argument that a crisis is at hand; they have been singled out and harshly criticized when they have undercut the environmental cause — namely, the organization of action to redress environmental degradation, if not to rescue the globe from perceived catastrophe. In an astonishingly short time, the environmental movement managed to transform scientific evidence into attention-getting headlines, even at times apocalyptic scenarios.

Despite science's central role in detecting environmental degradation, the scientific community has not been the "good guy" in the environmental story. As practiced by the scientific establishment, science and technology have more often than not been portrayed as culprits, even primary targets of the movement. Indeed, with the assistance of "movement scientists," the counterexperts of the environmental movement, much of the activity of the environmental movement since the 1960s has been stark opposition to the technocratic emphasis on scientific decision making that has dominated environmental policy making, not to mention industrial society more generally.[2]

In this sense, science no longer represents the essence of "enlightenment." Now it is greeted with skepticism. Indeed, for some, science is a primary force to struggle against. Portrayed as oblivious to the social and cultural dimensions that give life its meaning and purpose, scientific rationality is depicted as a form of *social irrationality* (Fischer and Hajer 1999). Basic here is the argument that scientific technologies are out of control.

The environmental movement, as such, is in significant part a reaction to the quasi-religious faith in science that emerged in the West after World War II. Nothing has been more important to many in the movement than defying technocratic experts and managers as the ultimate arbiters of technological advance and environmental risk. One of the strengths of the movement has been its ability to build on the frustrations and rage of people who see their quality of life threatened by technological systems and perceive themselves as victims. The irreverence toward official versions of reality offered by scientists and technocrats has provided a powerful organizing tool generally. Environmentalists have portrayed these technocratic orientations as rooted in an elite strategy designed to undercut democratic governance.

These technocratic tendencies have involved more than the appearance of new scientific technologies and decision techniques. Many environmentalists have also extensively criticized authoritarian arguments that have often accompanied them, in particular the arguments of writers such as Hardin (1968), Ophuls (1977), Heilbroner (1974), and Bahro (1987). These writers have taken the position that the environmental problem is too serious and complex to be left to conventional political decision-making processes, democratic or otherwise.

In this view, the profligate ways of affluent Western societies must be transformed. Given the capitalist society's tendencies to celebrate "selfish hedonism," these writers see the inevitability of more centrally planned authoritarian political structures. We are seen to face two grim choices: continue with business as usual and attempt to adjust to life in an impoverished environment, or abandon democracy in favor of a powerfully centralized but ecologically sensitive Leviathan. Surviving the disaster ahead, they argue, will require more centralized systems of authority, increased government intervention, scientific planning, resource rationing, population control, and authoritarian political structures.[3]

The German philosopher Bahro, for example, calls for the introduction of an "Environmental Council" with overriding power to protect the environment. For Bahro, the council might be likened to the House of Lords in the British parliamentary system. Staffed by environmental experts, the council would set out a framework of principles of sustainability and ensure that economic policy remains within it. Although Bahro sees these "ecologist-kings" as compatible with the idea of an environmental democracy, others have denounced his vision as a form of ecological dictatorship, or "ecofascism."

Environmental Democracy

Most environmentalists, however, treat the idea of an ecologist-king with derision. The answer, they argue, must be *more* rather than *less* democracy. Their case for more democracy has advanced along two diverging lines, one pluralistic, the other radical. The pluralistic approach has its origins in the political struggles of the 1960s. Insofar as the environmental movement emerged in tandem with other movements of that period, such as the peace, consumer, and civil rights movements, it was influenced by the public interest politics that shaped these movements. Basic to the leftist politics of this period generally was the need to broaden the participation of interest groups, especially the underrepresented groups. Following the lead of the civil rights movement in particular, many environmental groups adopted grassroots mobilization and participation as primary goals. This led many to formulate the environmental problem in terms of interest groups struggling to ensure a more equitable distribution of environmental "goods" (e.g., urban amenities) and "bads" (e.g., pollution). The growth of public concern over environmental problems was thus widely interpreted as involving participatory and distributional issues — that is, issues concerning "who decides" and "who gets what, when, and how."

This view still prevails in many environmental circles. Taking issue with those who maintain that ecological limits raise serious questions about the efficacy of political democracy, for example, Paehlke (1995) argues that environmental protection will be achieved effectively only through continuing enhancement of democratic practices. Pluralist democracy is the only system capable of legitimately balancing basic environmental values — ecology, health, and sustainability — against first-order values such as social justice, economic prosperity, and national security. Such writers see democracy as our best hope for mobilizing a transition to environmental sustainability.

This pluralist approach, and the interest group politics inherent to it, is not without its advantages. Indeed, environmental politics has in large part emerged through the environmentally oriented public interest groups that have advanced the movement. It is difficult to find important policy changes that haven't been initiated and advanced by such environmental groups. In an extensive cross-national study, for example, Jaenicke (1996) shows that the capacity for interest group participation plays as great a role (or one even greater) in advancing environmental policy as government institutional arrangements. Furthermore, comparative studies of environmental groups in Canada and the United States indicate that they play the central role in interpreting and making available technical environmental information to citizens (Pierce et al. 1992).

But while interest groups are important, they are not to be confused with citizen participation. Although interest groups represent citizens, especially "public interest groups," they are themselves hierarchical organizations frequently quite removed from the citizens for whom they speak. Often the connection between these groups and their members is over time little more than a mailing list. Moreover, from sociological studies of these environmental groups, we learn that they tend to become more professionally oriented. As they begin to speak the languages of the professions, they gradually tend toward more cautious, if not relatively conservative, strategies. Such professionals, as grassroots movements are quick to point out, too often move from the environmental sector to the high-paying world of industry, serving as consultants in a relatively tight-knit world of professional advisers.

In the case of the American environmental movement, as Dowie (1995) shows, the Washington-based environmental groups have tended to develop an all too cozy relationship with the industries they set out to battle. As their leaders have moved from the streets and the courts to corporate boardrooms, a new style of politics shaped at the bargaining table has tended to bring compromises that close off progressive alternatives. In this respect, one of the critical issues in contemporary environmental politics in the United States concerns the degree to which the regulatory compromises negotiated by environmental interest groups necessarily represent the views of the majority of the citizens worried about environmental issues. In fact, the major environmental organizations, having joined the federal lobbying establishment, are widely seen by their grassroots critics as having become part of Washington's so-called Beltway politics, in significant part at the expense of more aggressive approaches to protecting the environment. From this perspective, these interest groups need to better avail themselves of the citizens' views they seek to represent.

In the early 1970s, responding to these and other limitations of contemporary politics, a more radical participatory strand emerged in the environmental movement. This strand took the form of a counterculture within the environmental movement more generally. The counterculture, as Theodore Roszak (1995) explains, emerged as a "new social movement" that raised challenging questions about the sustainability of techno-industrial society and its consumption-oriented way of life. Standing outside existing institutions, the movement has sought to make its case in a revitalized civil society and its public sphere (Eckersley 1992).

Environmental Social Movements and the Public Sphere

Understanding the origins and dynamics of social movements has proved to be a complex undertaking. Social movements represent a widely perceived but insufficiently understood phenomenon basic to political transformation in the postwar West, in particular the decline of class-based politics. The newer social movements differ from the labor movement not only in terms of their memberships but in terms of their organizational structures, methods of political action, and political aims. In contrast to the dominant forms of social democratic politics, primarily concerned with the welfare state and the kinds of compensation it can provide, the politics of social movements emphasizes defending and restoring endangered ways of life.

Rather than operating within conventional politics, new environmental social movements — such as the women's movement and the student and peace movements — have advanced a fundamental socio-cultural critique of the established order. Offe (1985) has characterized this orientation as a radical "metapolitics" aimed at a critical evaluation of institutional assumptions of liberal-capitalistic systems and the interest group politics basic to their governance. New social movements, as such, are geared to new lines of societal cleavage — new conflicts, modes of action, and values. Emphasizing the failure of conventional politics to deal with these conflicts, as Melucci (1994) explains it, this metapolitics concerns itself with issues that established parties are unable to confront. Focusing as much or more on the conditions of political action, rather than action itself, these new struggles focus on the knowledge and skills that individual citizens and groups need to shape the conditions of both their personal lives and political actions. Toward this end, such movements emphasize information that "extends, in practice, from relatively narrow demands for the right of citizens' access to practical facts (such as the location of missile deployment sites or the extent of ecological damage caused by toxic waste dumps) to broader debates over symbolic resources, such as the challenge of the women's movement to the exploitation of women in pornography or advertising" (Breyman 1998, 23).

For environmental social movements, this metapolitics has been grounded in the critique of industrial growth. Whereas the "old paradigm" of postwar industrial politics fostered a societal consensus based on continual economic growth, the steady expansion of the state, and materialist values, the "new paradigm" launched by the progressive factions of the environmental movement confronts the presuppositions and costs of the dominant strategy. What is needed, according to these environmental theorists, is a more thorough reexamination of the basic axioms of liberal capitalist society, such as private property, limited government, interest group politics, consumerism, and market freedom. Toward this end, theorists have brought the very notion of material progress into question, as well as the social and psychological costs associated with the dominance of instrumental rationality. Included among these social-psychological costs are social and political alienation, loss of meaning, the coexistence of extremes of wealth and poverty, welfare dependency, dislocation of indigenous cultures, and the growth of an international urban monoculture with a concomitant reduction of diversity. For the new environmental social movement, Western industrial society's sanguine reliance on future "technological fixes" and better planning is increasingly recognized as the problem rather than the solution.

As a radical critique of techno-industrial society, new environmental movements contest the traditional doctrines of both liberal capitalism and Marxist socialism. Both ideologies are seen to be equally committed to expanding urban industrialism; both are equally wedded to a dogmatic acceptance of technology and science. Radical environmentalism challenges liberal capitalism's emphasis on regulatory rules, adjustive interventions, and representative-bureaucratic political institutions. For radical environmentalists, moreover, the pluralist conception of environmental democracy also fails. Although far superior to ecofascism, environmental democracy ultimately falls victim to the weakenesses of pluralism's emphasis on process over substance. In short, as these critics see it, there are no substantive values built into the system to ensure that the participants will decide to make saving the environment a top priority.

With regard to socialism, especially Marxist socialism, the movement disputes the usefulness of a narrow focus on social class conflict. For radical environmental theorists, like new social movement theorists generally, the older conflicts of industrial classes constitute an outmoded strategic political orientation. Not only have these basic class cleavages been institutionalized in formal procedures such as collective bargaining, the welfare state, and mass political parties, but they are seriously blurred by issues of gender and race. In the process, class has lost much of its critical leverage as an organizing concept.

Radical environmentalists reformulate the struggle in terms of both the apparent irrationalities of modernization and the newly developing forms of domination increasingly characteristic of postindustrial society. The costs of the economic and political rationalities of modern industrialism — such as pollution and bureaucratization — are seen as dispersing in time and space, affecting virtually every member of society beyond the conventional group criteria of class, gender, or race (Beck 1992). At the same time, processes of social control are no longer confined primarily to the workplace (as classical Marxism assumed). They now penetrate deeply into the realm of culture and social reproduction as well, giving rise to what has been described as a form of "cultural politics" (Fischer and Hajer 1999, 6–10).

For Habermas (1987, 394), the origins of radical social movements can be understood as a response to the "colonization of the social lifeworld." One of the primary sources of domination in modern society occurs as the increasingly bureaucratized systems of corporate capitalism and the bureaucratic state begin to expand their reach and "colonize" the social lifeworld. In short, the new conflicts are ignited less by questions of distribution than by matters having to do with the grammar of everyday social life. Environmental risks are certainly one of the primary cases in point.

As the intertwining of state and economy increases, along with the growth of complexity and bureaucratization of the social system, the means-ends rationality of the political economy becomes more and more prevalent in the family and the public sphere, supplanting the intersubjective communicative rationalities that define the lifeworld. The imperatives of money and power begin to redefine citizens as consumers, and private individuals as welfare state clients. As citizens become consumers, their political role shrinks, public discussion of political ends erodes, and voting or choosing among preset alternatives becomes the predominant citizen activity. Insofar as the state has already sided with capitalism, citizen participation can only concern the implementation of predetermined goals (Goldblatt 1996).

The political demands, organization, and targets of protest for the new social movements are thus based on the quality of life, individual self-realization, social identities, participation, and human rights. They express challenges to conventional definitions of social roles at the interface of system and lifeworld. Not only is consumerism rejected for its commodification of lifestyles, but unconventional political protest is employed to ridicule the thinness of the largely strategic character of conventional politics. Such protests express a rejection of mass-party representative democracy in favor of looser political organization and participatory democracy. New modes of self-help and participatory control replace bureaucratized institutions. Emphasizing this identity-oriented politics, such movements call for the creation of social spaces in which the lifeworld can operate according to its own normative dynamics.

The environmental movement, in this respect, represents a politics underwritten and motivated by moral values, or what is described as a "postmaterialist" value system. A core of young middle-class, well-educated men and women, less attracted to material values than their parents, experience those "pathologies" most acutely and are most sensitized to their existence. Indeed, their capacity to perceive the colonization process stems in significant part from their response to ecological problems. The physical interest in survival opens participants to the recognition of the destructive impact of strategic action systems now extending into the lifeworld.

One manifestation of an overly rationalized lifeworld is an excessive complexity that makes it difficult for people to grasp their own situation and thus to immediately recognize the material destruction of their natural and urban environments. So, too, distortions of the communicative infrastructure lead to fragmented consciousness, the loss of totalizing understandings of the world, and the bureaucratization of the will-forming process in the public sphere. For the members of this movement, the issue is how to escape the fragmented consciousness of the lifeworld that makes it difficult for people to see the whole.

The solution, in this view, lies in strategies situated beyond the reach of the dominant institutions. For this reason, the movement's radical theorists argue that the dominance of the "instrumental complex" of state and economy can be overcome only from outside the system. New environmental social movements have thus sought non- or anti-institutional strategies. Insofar as this metapolitics is aimed primarily at the cultural rather than economic or political foundations of society, they have chosen to focus on revitalizing civil society and the public sphere (Eder 1996). Offe (1985) has put it this way: "The politics of new social movements . . . seeks to politicize the institutions of civil society in ways that are not constrained by the channels of representative-bureaucratic political institutions, and thereby to reconstitute a civil society that is no longer dependent upon ever more regulation, control, and intervention." Basic to this task has been an effort to repoliticize the public sphere, defined as the social space between civil society and the state (Dryzek 1996). It is nothing less than an effort to reconstitute democracy in the postliberal period of the late twentieth century.

Environmentalism as Cultural Politics

From this perspective, a growing number of environmental thinkers have identified new cultural opportunities in what had hitherto been pessimistically approached by the earlier environmentalists as a dire crisis with a limited range of options (Fischer and Hajer 1999). This new breed of ecopolitical theorists — ecofeminists in particular — have drawn out what they see as an emancipatory potential within the ecological critique of industrialism. Such a project entails much more than a simple reassertion of the modern emancipatory idea of human autonomy or self-determination. It also calls for a reevaluation of the foundations of, and the conditions for, human autonomy or self-determination. Beyond an arena for pursuing more effective environmental decisions, a democratic public sphere is emphasized as essential to effective citizenship development. For such theorists, the revitalization of the civil sphere also represents a way of integrating — both theoretically and practically — the concerns of the environmental movement with other new social movements, particularly the feminist, peace, and Third World development movements. Together, they could struggle to build the new consciousness needed for bringing about fundamental changes in our ways of life.

One of the most significant practical expressions of the emancipatory approach to environmental politics today is found in the environmental justice movement, a radical environmental populist movement that has emerged within the environmental movement more generally to confront the toxic waste crisis in the United States (Bullard 1993; R. Moore and Head 1993). Originating in the discovery of toxic waste sites and the community-based politics of NIMBY (not in my backyard) of the past decade, the movement rapidly developed as a response to the racial and social class disparities in environmental and occupational health. Pointing to the disproportionate incidence of environmental health disorders in low-income and working-class communities — particularly those of African Americans, Latinos, Asian Americans, and other persons of color — the movement has sought to articulate the experiences of millions of low-income and working-class persons living near hazardous waste sites in the United States. In recent years, the movement has further expanded the potential size of its constituency by focusing its activities on the problems of toxic chemicals in all of their forms.

Basic to the struggles of environmental justice activists — a majority of whom are women — is an effort to produce the conditions for social and environmental change, locally and nationwide, by reinventing socioeconomic terms and definitions, constructs of gender, race, and class politics, notions of social movement history, forms of leadership, and strategies for coalitions (Schlosberg 1999). The movement has, in this way, become an important component of "cultural politics" more generally.

Focusing on the particular way of life of a people or a group, including the signifying system through which it is communicated, reproduced, experienced, and explored, cultural politics seeks to politicize the very foundations of policy-making institutions (Hofrichter 1993). Culture is a set of material practices that constitute the meanings, values, and identities of a social order. For example, in the environmental justice movement, this has meant, among other things, identifying the ways in which our understandings of environmental issues are built on patterns of racism fixed in the interpretive contexts that shape institutional decision making. In the process, the movement's theorists seek to show that institutions and their values not only are part of the environmental problem but are often *the* problem (Harvey 1999).

Fundamental to the movement's coalitional activities is an attempt to bring together a wide range of other social movements, including labor unions, civil rights groups, religious and interfaith activists, tenants rights groups, gay and lesbian organizations, antiwar and anti–nuclear power movements, citizens groups and other environmentalists. With these groups, they have sought to find common ground around the interrelated problems of toxic and hazardous wastes, air and water contamination, industrial pollution, workplace safety, and the critique of the basic values that undergird industrial society.

Activists in the environmental justice movement emphasize the transformative and empowering impact of such activities on individuals. Not only does the movement challenge the dominant discourse of environmentalism, but it seeks to produce new constructs for environmental empowerment and action. No premise is more fundamental than the idea that people are an integral part of what should be understood as "the environment." Whereas more conventional efforts have mainly focused on the environment in physical terms — as external to the nature of human interaction — the environmental justice movement seeks to explain environmental degradation in terms of links between physical degradation and the other pressing social and political problems, especially as they manifest themselves in the daily realities and conditions of people's lives. In this respect, environmental justice is about sociocultural transformation broadly understood. It is directed toward meeting pressing human needs and enhancing the overall quality of life in many spheres — including health care, housing, species preservation, food, the economy, and democracy. It is about the use of resources in a sustainable world. For these reasons, numerous writers have held out the environmental justice movement as the truly progressive environmental force in the United States with the potential of revitalizing the environmental movement as a whole (Dowie 1995).

One of the most innovative features of the environmental justice movement's efforts to empower citizens and thus revivify democracy has been the effort to help local citizens understand their own needs and interests. In the case of toxic wastes, this effort has almost always involved confronting and coming to terms with scientific information about risk and exposure. Rather than merely accepting information provided by scientists and other technical experts — often engaged by industry or government to assure citizens that they should have little worry about toxic exposures — the movement assists communities in a variety of ways to collect and interpret their own information. In fact, in conjunction with movement-oriented scientists, this has given rise to an emerging form of lay expertise. It involves a method and practice of participatory research that goes considerable distance toward democratizing the otherwise hierarchical relationship between scientists and the communities they attempt to assist. Of particular importance, in this respect, has been the practice of "popular epidemiology," which I examine in chapter 8.

Cultural Politics as Ideology?

The Politics of Rationality

But what does this cultural metapolitics have to do with the practical world of regulating risks? Is it not merely an ideological critique of the modern industrial world and the way of life to which it has given rise? This is certainly the argument of many critics of the environmental movement (Rubin 1994; D. Lee 1990). Indeed, it is not uncommon for such writers to accuse environmentalists of merely engaging in an ideological critique of capitalism, this time in the name of cleaning up the environment. Socialism has collapsed, they argue, and those who hate capitalism have merely found the environment to be a new issue through which they can reformulate and further advance their critique (Boloch and Lyons 1993, 8; Baily 1993). Typically, such critics also contend that environmentalists either fail to understand science or misuse it for ideological purposes (D. Lee 1990; Fumento 1993). The movement, in this view, has discovered that it can play on the public's irrational fears through what critics have dubbed "the environmental crisis of the month" strategy (Bast et al. 1994, vii–viii). Environmentalists, it is further argued, should give up their ideological scare tactics and turn to "sound reason" for guidance. As Bast, Hill, and Rue (1994, 268) put it, "The environmental movement often confused scientific and economic problems with moral issues. Pollution, for example, was considered evil, while unspoiled nature was good." The "polluters — usually faceless corporations — were portrayed as villains, while popular reformers were treated as selfless crusaders." Making compromises "was usually seen as surrendering principles, rather than as a necessary step toward achieving goals." Calling for a new "Eco-sanity," such critics appeal to environmentalists to drop their moralistic rhetoric and to substitute it with rigorous science and good judgment.

Much of this concern about the movement's politics has emerged in response to the appearance of the NIMBY phenomenon that has plagued industrial countries over the past two decades. Even if citizens are willing to accept potentially hazardous technologies, such as nuclear power plants or waste incinerators, they would prefer that such installations not be located in their neighborhoods (Piller 1991). Worse yet is NIABY (not in *anyone's* backyard). Some environmentalists have labored to parlay NIMBY into this more radical position — namely, that hazardous technologies shouldn't be sited at all. And not without some success. For example, a combination of NIMBY and NIABY account in large part for the fact that no new nuclear power plants and few toxic waste incinerators have been built in the United States since the late 1970s.

Fundamental to these conflicts has been the refusal of citizens to accept the technical safety findings of the risk experts. Frustrated by the unwillingness of citizens to accept their assessments, experts and industry officials have often been quick to declare the public to be "irrational," unable to understand or comprehend technical data purportedly showing the low probability of an accident. In the throes of "ignorance," the public is seen to fall back on unfounded fears. Many environmental critics are fond of pointing out that citizens more readily accept levels of risk associated with, for example, smoking cigarettes, than living near an incinerator.

But how irrational is this behavior of the public? Are these fears as ill founded as the critics of the movement would have us believe? Taken from another perspective, we can begin to see that it is less a matter of ignorance than a different way of thinking about risks, one related more to a cultural logic than to technical calculations. Such an interpretation, moreover, suggests that what might be thought of as social or cultural reason is not an inferior mode of thought when compared to technical reason, as risk experts have tried to convince us; rather, it only relates to a different part of the problem that the experts have neglected. For writers such as Plough and Krimsky (1987), for example, "cultural rationality" emerges as a necessary and complementary mode that technical experts have failed to understand. In this view, the argument that the environmental movement and the public more generally are "irrational" rests on a false or limited understanding of both the nature of risk and the community decision-making process.

In the next chapter, I delve in more detail into the connections between the "cultural politics" of the environmental movement and the concept of cultural rationality that has emerged in the risk literature, a relationship that has thus far failed to receive sufficient attention. I will argue that cultural rationality and its emphasis on case-specific social processes not only is fundamental to citizen decision making but is the essence of rationality in certain types of decisions. Such rationality has to become an integral part of risk assessment. Thus the challenge ahead is not just more science but rather how to better understand the interactions between science and ideology — technical facts and cultural values — and most importantly how to integrate them systematically in a more comprehensive analysis.

7. Not in My Backyard

Risk Assessment and the Politics of

Cultural Rationality

In an area so complicated, so crucial and so fraught with
controversy . . . nonexperts like you and me can probably never bone
up enough to make our own independent prognoses based on sheer
data. We rely on experts to help us see what the numbers, the
models, the inputs and parameters actually mean — and, since the
experts disagree, we rely on our own decisions about which of
them we should trust. — David Quammen

During the late 1970s and 1980s, news reports of oil spills, nuclear
disaster at Chernobyl, near disaster at Three Mile Island, pesticides in
the food chain, and DDT damage to wildlife have frightened people
around the world (Piller 1991). The result has been a widespread distrust of industry and a collective fear of all chemical processing facilities, both of which have given considerable impetus to the environmental movement. As the public has become increasingly aware of the
extent to which chemicals pollute the environment, the result has been
a new anxiety often described as "chemophobia." Polls show that
citizens are more concerned about the presence of toxic wastes than
any other environmental problem, although the EPA maintains that
toxic waste is not the most severe threat. Problems such as the ozone
hole and the greenhouse effect are said to present far greater risks.

The NIMBY Phenomenon

As introduced in the previous chapter, one of the clearest manifesta-tions of this anxiety and distrust has come to be called the NIMBY syndrome. Much discussed in both the academic and popular presses, NIMBY — or "not in my backyard" — is now blamed as a major stum-bling block for solving a growing number of environmental problems. As one leading journal put it: "Once the public went along with every-thing: now it opposes everything."

NIMBY covers a wide range of activities. As Piller (1991, 12) ex-plains it, "Whether the matter is health, peace of mind, or protection of property values, few Americans (activists or not) care to live beside chemical-waste dumps, airports, petrochemical refineries, nuclear power plants, or other standard features of a modern industrial so-ciety." But while most often used as a term to designate opposition in general, NIMBY can better be understood as a description of a specific type of opposition.

Basically, NIMBY reflects a public attitude that seems to be almost self-contradictory: namely, that people feel it is desirable to site a par-ticular type of facility somewhere as long as it is not where *they* live. Moreover, NIMBY has spread from one policy area after another: land-fills, prisons, power plants (nuclear or otherwise), industrial parks, housing for the homeless, treatment facilities for drug addicts, and hazardous waste facilities (1992).[1]

There is thus no single NIMBY goal; activists are quite pluralistic in their strategies and objectives. Yet these varied groups are united in a number of basic ways. "Regardless of their demographic traits, NIMBY battles share common characteristics: Nearly all begin with the frustrated rage and fear of people who perceive themselves as victims and who see their quality of life threatened" (Piller 1991, 12). Highly focused on protecting their home environments, NIMBY activ-ists have wasted little time at becoming skilled at petition drives, polit-ical lobbying, street confrontations, and legal proceedings.

If their frustrated rage and anxiety are the most general characteris-tics that unite these groups, the most specific is their defiance of ex-perts and technocrats as the ultimate arbiters of technological risk and change. The zeal of NIMBY groups often, in this regard, takes on an aura of "proselytic self-righteousness." In fact, some have likened NIMBY activists to other moral and religious movements that have gained large followings by advancing what can be described as "a spiritual critique of medical or scientific teachings and practices." Piller (1991, 12) comments: "Although the link between NIMBY groups and right-wing religious movements are otherwise tenuous, they share irreverence for official versions of reality offered by scientists and technocrats." In his view, it can ironically be argued that the politics of NIMBY is "partly a reaction to the effects of the quasi-religious faith in science that emerged in this country following the Second World War." Indeed, the phenomenon is a manifestation of the end of the technological optimism that has long defined the "American Century" (Hughes 1991).

NIMBY and Risk Assessment

Thanks to NIMBY, community resistance to the siting of risky facilities can now be described as a "full-scale public malady," a kind of malignant social "syndrome" (Portney 1991, 10–11). Writers speak of "policy gridlock" and "policy stalemate." In the case of hazardous waste treatment facilities, for example, sitings have more or less come to a halt during the past decade.

The official response of government and industry to NIMBY and the fear of toxic risks, as we saw in chapter 5, has been to submit the dangers to a risk assessment (Fischer 1990; Wynne 1987). That is, formal policy analysis has been used in an attempt to decide the issue "rationally" by focusing the risk debate on technical factors. Specifically, the purpose of this orientation has been to shift the political discourse to a search for what has been termed "acceptable risk." The supporters of the modern techno-industrial complex argue that risk must be seen as a mixed phenomenon, always producing both danger and opportunity. Too often, they argue, the debate revolves purely around potential dangers (frequently centering on high-impact accidents with low probability—e.g., nuclear meltdowns or runaway genetic mutations). The approach is grounded in the view that technological dangers have been grossly exaggerated, especially by environmentalists who have a vested interest in exploiting the public's fears. The result, it is argued, is a high degree of ignorance in the general public about technological risks (Wildavsky 1988). The classic illustration is the layperson who tends to worry a great deal about the safety of air travel but thinks nothing of driving a car to the airport, a trip that statistics indicate to be a much more dangerous undertaking.[2]

The goal of risk assessment is to supply the public with more "objective" information about the levels of risks. That is, the "irrationality" of contemporary political arguments is to be countered with "rationally" demonstrable scientific data. The solution is to provide more information — standardized scientific information — to offset the misperceptions and distorted understandings plaguing uninformed thinkers, particularly the proverbial "man on the street." The objective of this new line of research has been to figure out how to more persuasively convey the technical data to override the "irrational fears" of ordinary citizens.

But these risk methodologies haven't typically achieved their purpose. Indeed, they have not only tended to heighten the conflict but often become the very bones of contention. The typical conflict involves a government study that shows a low level of risk and a public (or at least the community groups most directly involved) that is adamantly opposed to accepting the findings. In short, people have remained unswayed by risk assessments, and the result has been a near halt in the siting of hazardous waste facilities. This reluctance to accept such risk analyses has compelled environmental officials to call community responses "irrational." NIMBY groups have either rejected the experts altogether or sought to commission their own analyses. The result has become an impending crisis: in the face of growing mounds of dangerous waste, there is no place to treat or store it.

NIMBY as a "Wicked" Problem

The failure of risk assessment and risk communication to solve — or at least assuage — the problem of NIMBY has led to a great deal of investigation into the nature of the conflict. Research has tried to uncover why citizens are so averse to accepting the technical findings of the risk assessors. Although a full explanation of this phenomenon defies simple explanation, risk assessment's dilemma is rooted in its techni-

cal orientation to risk. The difficulty is lodged in the treatment of risk
as a relatively structured problem that can be approached in a relatively uncomplicated and straightforward quantitative fashion. This
is not to suggest that this kind of analysis is uncomplicated. It is only
to note that in a standard risk assessment, the "problem" is taken as
given: we need only to find the answer.

A closer look at the dilemma of risk suggests that it can better be
understood as a *poorly* structured problem, in particular one that has
come to be identified as "intractable" or "wicked." Wicked problems
are those in which we not only don't know the solution but are not
even sure what the problem is. Whereas traditional problems in government — such as paving the roads, connecting the sewers, or administering public schools — can be attacked successfully with common
sense and ingenuity, and thus have been described as relatively "tame"
and "malleable," they have in more recent years given way to a different class of problems with only temporary or uncertain solutions.
Involving issues such as drug addiction, biodiversity, AIDS, the homeless, or the siting of incinerators, the solutions to these problems tend
to be "tricky," run in "vicious" circles, and spread like a "malignant"
growth (Harmon and Mayer 1986, 9; Rittel and Webber 1973, 160).

So-called tame problems are solvable because they can be easily
defined and separated from other problems. This is not to underestimate such problems by implying they are easily dealt with. Rather, it is
only to identify their common feature: namely, that they are largely
technical in nature. In sharp contrast, "wicked" problems lend themselves to no unambiguous or conclusive formulations and thus have
no clear-cut criteria by which their resolution can be judged. As Harmon and Mayer (1986, 9) put it, "The choice of a definition of such a
problem, in fact, typically determines its 'solution.' "

Focusing on the difficulties involved in defining a wicked problem,
Hoppe and Peterse (1993, 25) explain that "in every problem, two
heterogeneous 'elements' are linked to each other: normative criteria
(objectives, standards, rules, etc.) and empirical situations or conditions." A problem, in this respect, is not a given fact, or something
from the outside world. It is itself a social construct. A policy problem
thus involves a gap between norms and an empirical situation. Neither the standard nor the situation is an objective datum exterior to
the social actors. Both are social constructs based on social actions

and judgments. We can only say that there are normative standards that command more or less consensus, and that there are situations about which there are greater or lesser amounts of certain knowledge.

The problem of siting hazardous facilities confronts risk assessment with just such a problem. Accounts of the conflicts surrounding NIMBY are almost invariably described with terms such as "undisciplined," "uncontrollable," "recalcitrant," and "unmanageable." Often overlooked is the fact that such terms constitute as much a description of the risk problem itself as they do the politics of NIMBY. That is, not only do NIMBY activists typically reject the technical findings of the risk analysts; they question the very way the problem is understood and defined. In some cases, for example, the question is as much about how issues of local land use are decided as it is about the safety of a particular facility. In such situations, the use of quantitative risk assessment is at a severe disadvantage. Insofar as the issues concern more the norms and standards on which the quantitative analysis rests, rather than the findings per se, the kinds of normative discourse and analysis needed to address the conflicting issues lie outside of the repertoire of empirical methods. Empirical analysis has, in short, no way of establishing the validity of its assessment criteria independently of the assessments of the local actors. Put in another way, the uncertainty of the empirical situation renders its analytical findings open to alternative interpretations. For quantitative risk assessment, based as it is on the concept of objective data, such problems have proven to be something of an intellectual embarrassment.

The Participatory Alternative

What, then, is the alternative? Many environmentalists argue that the answer is more participation (Thornton 1991; Paehlke 1990; Kann 1986). In their view, it is the right of citizens to participate in, and collectively decide, questions about technology and the environment. The goal should be to remove the privileged status of the experts employed by industry and government and to provide citizens with opportunities to contribute to decisions about environmental issues that affect their own interests.

Although most mainstream economic and political leaders take the participatory perspective of environmentalists to be mere ideology, more recent experiences have begun to suggest that there is more to the argument than might first appear to be the case. Siting efforts in Canada, for example, now appear to validate — at least provision-ally — the environmentalists' emphasis on participation in dealing with these "wicked" problems. These experiences suggest that the solution to the doubt and mistrust surrounding technical risk assess-ments is a more open set of communicative relationships, in particular something beyond the technocratic approach to "risk communica-tion" introduced in chapter 4.

A variety of efforts, from experimental to practical, show the value of open participatory communications. On the experimental side, Elliot (1984) has used quasi-experimental game simulations to show that open communications among all the participants can signifi-cantly increase the possibilities of coming to agreements. Employing a diverse sample of participants in two communities — including public officials, business groups, environmentalists, community leaders, and landowners — Elliot found that public officials and technical experts mainly focused on the technical aspects of a siting decision, whereas community members tended to be preoccupied with detecting en-vironmental health risks associated with the management of the facil-ity. For members of the community, the crucial issues are the kinds of safety procedures and processes needed to mitigate or reverse poten-tial dangers. Most important, Elliot's results showed that community participants were even inclined to approve of a less technically sophis-ticated facility if its managers sufficiently stressed effective detection and quick responses to both immediate and future problems. Such findings reveal citizens to be not necessarily hostile to technical data. They might well be willing to accept such data if they are presented and discussed in an open democratic process. This has led a number of writers to conclude that open participatory decisions are the only way around NIMBY.

Such experimental findings are now corroborated by practical ef-forts to employ extensive participation in the siting of actual facilities. Most important are a set of experiences in the Canadian province of Alberta. Faced with the NIMBY syndrome, the regional government decided to confront openly and squarely the community opposition to a plan to build a new incinerator (Rabe 1991, 1992; Paehlke and

Torgerson 1992). By setting up an open and democratic participatory process, the government did the undoable: it successfully managed to site, build, and operate the single major new incineration facility in North America in more than a decade.[3] Working together, government, industry, and local groups in Swan Hills, Alberta, devised a participatory process through which the conflictual issues of siting were transformed. Through the process, all major stakeholders ended up preferring negotiation to conflict; all came to see benefits from cooperation. Gone were the winners and losers that have typically framed the "zero-sum" politics of NIMBY.

Participation in the project was built into the decision process from the beginning, commencing with a local plebiscite on the acceptability of the siting decision. After the plebiscite, in which the citizens signaled their willingness to consider accepting the plant, the regional government supplied the Swan Hills community with funds to hire its own experts and consultants and organized extensive public meetings to discuss with community members and their consultants the nature of the plant and its consequences. Once the site was accepted, the government provided the community with additional monies to offset the extra burdens to the local infrastructure and to hire its own expert advice.

The community group in Swan Hills used the money to establish a local committee to organize seminars and meetings for community residents regarding hazardous waste treatment. Meeting on a monthly basis, the local committee sought both to provide facility managers with a source of information regarding community attitudes and ideas and to review reports that monitored the operations of the plant. The reports were translated from technical language into an easily understood format. The government also provided the community with money to hire a permanent consultant to assist them in monitoring the facility's operations.

Although the Canadian experience might not work everywhere, it has shown that a democratically inspired discourse can be constructed toward a positive end for the most complex and fear-invoking type of facility.[4] As Mazmanian and Morell (1993, 28) have put it, the Canadian experience suggests "that to increase the likelihood of siting good facilities in good locations, long-term oversight arrangements that provide for greater community involvement, power sharing, and risk sharing will be necessary." Contrary to technocratic assumptions, citizens' participation can prove to be anything but irrational.[5]

Even more significant, a careful look shows that "rational" here tends to have a different meaning from the one assigned to it by risk analysts. In the remainder of the chapter, I argue that community members bring to the decision process a different kind of rationality, one that has largely escaped the analytical nets of technical experts. Furthermore, the key to success in overcoming NIMBY is to find a way to build this cultural rationality into the decision process.

Technical Knowledge in Cultural Context

While risk assessment has been employed — albeit unsuccessfully — to circumvent the "irrationalities" of citizen decision making, additional research into the question of why communities have so adamantly rejected the advice of the experts offers quite a different perspective. Whereas risk experts have portrayed the environmental movement and the public more generally as incapable of digesting technical findings, and thus as susceptible to irrational fears, Plough and Krimsky (1987) show that such conclusions rest on a limited understanding of the nature of risk and the community decision-making process.

In their work on environmental risk assessment, Krimsky and Plough, as indicated in chapter 6, contrast the concept of technical rationality with the idea of "cultural rationality." "Technical rationality," they explain, is a mind-set that puts its faith in empirical evidence and the scientific method; it relies on expert judgments in making policy decisions. Emphasizing logical consistency and universality of findings, it focuses attention in public decision making on quantifiable impacts. "Cultural rationality," in contrast, is geared to — or at least gives equal weight to — personal and familiar experiences rather than depersonalized technical calculations. Focusing on the opinions of traditional social and peer groups, cultural rationality takes unanticipated consequences to be fully relevant to near-term decision making and trusts process over outcomes. Beyond statistical probabilities and risk-benefit ratios, public risk perception is understood through a distinctive form of rationality, one that is shaped by the circumstances under which the risk is identified and publicized, the standing or place of the individual in his or her community, and the social values of the community as a whole. Cultural rationality, in this respect, can be understood as the rationality of the social lifeworld. It is concerned with the impacts, intrusions, and implications of a particular event or phenomenon on the social relations that constitute that world. Such concerns, as we saw in chapter 5, are the stuff on which the environmental movement was built.

Chernobyl and the Local Assessment of Risk

The nature of the critical reflection of local citizens on matters pertaining to risk has been analyzed by Brian Wynne (1992) in his study of the reactions of northwestern English sheep farmers to scientific reports about the safety of radioactive contamination caused by the Chernobyl nuclear fallout in 1986. Subjected to administrative restrictions on sheep grazing and commercial sales, the farmers interacted for two years with government scientists responsible for both the restrictions and the official governmental position on the behavior of the radioactivity.

Numerous examples illustrated the failure of the experts to examine the particular geographical contexts, local knowledges, and practices of the farmers they were advising. In one case, for instance, the experts suggested that the farmers should avoid allowing their sheep to graze on the higher, more contaminated parts of particular hills. For the farmer this was absurd. As one farmer expressed it, the experts "don't understand our way of life. They just think that you stand at the bottom of the hill and wave a handkerchief and all the sheep come running." When they tell you "things like that it makes your hair stand on end. You wonder, what the hell are these blokes talking about?" (Wynne 1996, 66)

Drawing on his research as a whole, Wynne identifies the following kinds of questions and criteria that emerged in the course of the farmers' lay assessments of the scientific judgments put forth by the government experts. Although based on a particular experience, he suggests that they can approximate a more general set of "criteria by which laypeople rationally judge the credibility and boundaries of authority of experts of knowledge."

Wynne found that the farmers asked questions about the nature of the scientific predictions made by scientific experts. Will the predictions work or fail? Furthermore, they asked if the scientific experts had paid attention to other available types of knowledges? For instance, did the radiation experts seek out the sheep farmers' local knowledge about conditions of the plots on which the animals graze? Similarly, they asked the same thing about the specific scientific practices involved. Did the practices take into account the kinds of local information the farmers have gained through long experience on these lands?

In addition, they raised questions about both the content and form of the experts' knowledge. Are differing levels of uncertainty expressed in ways that are readily comparable?

What happens when the scientists are criticized? Do they acknowledge the existence of other types of experts? Do they admit errors or failures of omission?

Beyond these technically oriented questions they also raised questions related to cultural rationality: What are the institutional and social affiliations of the particular scientific experts? Are there obvious concerns or worry about social bias? Do they have credible records of openness and candor? Can they be considered trustworthy? And finally, are there other kinds of lay experiences that have a "spillover" effect on the current decision-problem, such as knowledge of previous nuclear accidents that might influence the farmer's or scientists thinking about the fallout on the sheep grazing lands.

What do we know about the ordinary citizen's approach to risk? For the layperson, the concept of risk is understood as much in terms of qualitative, affective characteristics as it is in terms of quantitative relationships. Psychological research into the perception of risk shows that citizens' understandings of risk are made up of a rich, multifaceted perspective that includes perhaps nineteen or twenty affective characteristics (Covello 1993; Slovic et al. 1979). Such characteristics tend to be expressed in terms of basic dichotomies: the voluntariness versus involuntariness of the risky action; its familiarity or unfamiliarity to the people involved; the immediacy or delay of the potentially risky effects; whether the risk is produced by natural or artificial forces; who controls the risk (the individual at risk or someone else); the visible or invisible nature of the benefits; and so on. According to this research, the more involuntary, unfamiliar, unfair, or invisible the risk, the more likely citizens will be to oppose it (Kasperson and Stallen 1991).

Focusing on how laypersons cognitively process uncertain information, social psychological research shows the ways in which citizens draw on past experiences in making assessments. Given the complexity of most policy issues, especially technological ones, citizens tend to fill knowledge gaps with information about social process, or what has been called the "social process theory" of cognition (Hill 1992).

Not all people, of course, have the same experiences. It is possible, in this regard, to think of a continuum across which people with different levels of experience can be distributed. Individuals, such as public administrators or political activists, who have considerable experience with a particular issue or problem, develop relatively abstract and well-integrated knowledge structures that actively guide their perceptions and expectations in future decisions. These "schemas" inform such individuals or groups about how events are expected to unfold, as well as how particular people should act in given sets of circumstances (Conover 1984; Fiske and Taylor 1984). They also explain how substantive issues in a particular area of politics interrelate or how decision-making procedures should operate. Members of the lay public who spend much less time dealing with and thinking about policy issues invariably hold less-developed schemas. Their ability to perceive and analyze the various dimensions of comparable issues, as a result, is necessarily far more limited, often giving the impression that such people are uninformed. What the research shows, however, is that in such situations, citizens rely more heavily on *procedural* than on substantive schemas. Citizens turn to these often well developed generalized procedural schemas that can be applied to a range of different situations, from political decision making to committee work in the office.

Perhaps the most important effort to test this social process theory of cognition in the case of a policy decision is Hill's (1992) sophisticated empirical study of the role of past experiences in citizens' assessments of a nuclear power plant in California. Hill's research documents the ways in which technical experts and administrative policy specialists have largely misunderstood the thinking of the lay public. Not surprisingly, he found the laypersons to have nothing of the technical understanding of the scientific experts involved with the plant. At the same time, however, he found the citizens' cognitive processes to be far from confused or uninformed. Rather, citizens conformed to the social or cultural process model focusing on case-specific contextual information. Whereas technical experts and nuclear power managers portrayed the new plant as providing local citizens with additional sources of electrical power coupled with a lower local tax base, local residents focused attention on safety procedures. The principal concern that opened a lengthy public debate about the siting of the plant was that it had inadvertently been located near an earthquake fault line and that the engineers had failed to properly equip the plant to withstand a sizable quake. Although technical in nature, this question wedged the debate open to a wide range of social and political questions about the engineers associated with this failure. Hill's findings led him to conclude that the laypersons' knowledge was not just different from the technical knowledge of the experts but in fact a *complement* to the assessment methods of the nuclear experts and politicians. By judging how well the general engineering arguments in support of the plant applied to the specific substantive impacts of decisions in a particular local context, the lay public's emphasis on case-specific social processes effectively counterbalanced the technical expertise of the evaluators (Hill 1992, 26).

The turn to cultural rationality and its emphasis on social process is most apparent in the case of uncertain data. Uncertainty opens the door for competing interests to emphasize different interpretations of the findings. "Wicked" problems such as NIMBY, moreover, raise normative as well as empirical uncertainty. The question of how to define the situation is as problematic as the question of what to do about it. Competing definitions emerge from multiple, often conflicting perspectives. Normatively, in such cases, politicians and activists advance counterarguments about the nature or definition of the problem. Empirically, each side engages in what I have previously described as the politics of expertise, employing the same or similar data to suit their own purposes.

And where does this leave the public? Consider the empirical dimension of the problem. If two experts stand before an audience of citizens and argue over the empirical reliability of a given set of statistics, what basis does the citizen have for judging the competing empirical claims? In this situation, citizens are forced to rely more on a sociocultural assessment of the factors surrounding a decision. And not without good reason. Although scientific experts continue to maintain that their research is "value neutral," the limits of this view become especially apparent once they introduce their technical findings into the sociopolitical world of competing interests. In the absence of empirical agreement, there is every reason to believe that interested parties will strongly assert themselves, advocating the findings that best suit their interests. In such cases, at least in the immediate situation, there is nothing science can do to mediate between such claims. One can call for more research, but as experience shows, there is little guarantee that further research will bring either certainty or timely results in a particular conflict.

Reliance on cultural rationality is especially strong when there is reason to believe in the possibility of deception or manipulation, which has often proved to be the case in environmental politics. In a world of industrial giants with vastly disproportionate power and influence compared to that of local communities, it comes as no surprise that citizens tend to be wary of the kinds of distorted communications to which such asymmetrical relations can give rise. When citizens have compelling reasons to suspect that a risk assessment is superficial or false, they can only turn to their own cultural logic and examine the results in terms of previous social experiences. Turning away from the empirical studies, they ask, What are our previous experiences with these people? Is there reason to believe we can trust them? Why are they telling us this? (Perhaps even, Why don't they look us in the eye when they tell it?) Such questions are especially pertinent when crucial decisions are made by distant, anonymous, and hierarchical organizations. Citizens want to know how conclusions were reached, whose interests are at stake, if the process reflects a hidden agenda, who is responsible, what protection they have if something goes wrong, and so on. If they believe that the project engineers and managers either don't know what they are talking about or are willing to lie or deceive to serve the purposes of their company, workers or citizens will obviously reject the risk assessment statistics put forth by the company. For example, if they have experiences that suggest they should be highly distrustful of particular plant company representatives or plant managers, such information will tend to override the data itself. From the perspective of cultural rationality, to act otherwise would be irrational.

Most fundamentally, cultural rationality, as an informal logic deduced from past social experiences, tells citizens whom they can trust and whom they can't. In chapter 3, we saw how citizens' and workers' understandings of large-scale technologies are shaped by the sociohistorical context in which they are embedded and experienced (Fischer 1991b). Technology is encountered as more than an assemblage of physical properties; it is experienced as an interplay between physical properties and institutional characteristics (Wynne 1987). As such, the ordinary social perceptions and assessments of technological risks by workers and citizens are rooted in their empirical social experiences with the technology's managerial decision structures as well as historically conditioned relationships, interpreted and passed along by members of their own groups and communities. The social relations of the workers and managers are pervaded by mistrust and hostility; the uncertainties of physical risks are amplified.

Trust is an essential category of modern sociocultural knowledge (Giddens 1990, 79–111). Expressed as confidence in some attribute or quality of a person, thing, or statement, trust helps us orient to one another; it serves as a basic social cement of group and societal integration. In modern societies, where institutions and practices are based on "abstracted" expert knowledges, trust takes on special importance. Because such knowledges are "disembedded" from the local contexts to which they are applied, people are left to trust in the validity of the knowledge and the competence of the expert who administers it (Giddens 1990). Surrounded by expert systems whose validities are said to be independent of time and space, we usually have little choice but to rely on the decisions of faceless authorities. For instance, such trust permits us to take the elevator to the top of a skyscraper, even though we know little about the principles of architectural design or building codes. Indeed, our entire existence is circumscribed by similar situations, whether we are flying in a plane, driving a car, eating food in a restaurant, or visiting a doctor.

However, in a society where the level of trust is low, cultural rationality will most likely caution citizens to be skeptical or resistant. Our sociocultural experiences are, in this way, factored into our interpretations of the experts' technical data on risk. Such data, after all, are not only a statement about the degree of danger we face but also a statement about the degree of danger in which another group has placed us. On this aspect of risk, the technical findings are silent.

While laypersons tend to rely heavily on sociocultural rationality, it is crucial to note that few people—whether laypersons or experts—act or think exclusively in one mode of rationality or the other. Such modes typically change with circumstances. For example, Sandman has demonstrated this phenomenon with a simple test. He asked experts to imagine themselves in situations in which they were not in control of the surrounding circumstances and to think of themselves as fathers rather than as engineers and businessmen. In such cases, the experts abandoned the technically rational model of decision making for the sociocultural rational mode of the citizen (cited in Hadden 1991). That is, the experts responded in their roles as citizens. For the experts as well, the evidence they were given was insufficient. When it came to protecting their own families, the matter of trust required knowing more about the social processes behind the reported evidence. The exercise demonstrates that cultural rationality is a different kind of knowledge that must be taken into account in any decision-making process.

At this point, we can recognize that the critics who argue that the environmental movement is grounded more in social critique or political ideology than in good science are not entirely wrong (Rubin 1994; Douglas and Wildavsky 1982). Without entering terminological disputes, the concept of sociocultural rationality bears a family resemblance to the concept of ideology, if it is not the same thing. As used here, it might even be understood as the deductive rationality of an ideological belief system. Insofar as citizens interpret risks from the perspective of sociocultural experiences, they do so within such a belief system. As deductive distillations of their experiences, these beliefs supply them with guidelines for action based on past experiences. Thus, in situations that are unclear, uncertain, or anxiety provoking, citizens are especially open or amenable to such appeals, and the environmental movement stands ready with ideological assistance.

What the critics fail to recognize or acknowledge, however, is that such a sociocultural perspective is inherent to the nature of the decision process. That is, in such situations there is no alternative but to seek out such normative guidance. Interpretations of how the social system works are precisely the kind of information that citizens need to help them link their own knowledge and experiences into meaningful understandings of a particular situation. As the turn to basic cultural orientations is in significant part a response to the fact that science cannot supply the needed answers, it is thus anything but irrational. Although critics portray the movement as merely appealing to the lesser instincts of the citizenry, their own call for more emphasis on science rather than ideology fails to grasp this point. Wittingly or unwittingly, science serves itself as an ideology.

Recognizing the role of ideology in environmental decision making, critics such as Rubin make the mistake of ascribing to it only a utopian interest in social change. In doing so, they fail to see that the ideology of environmentalism fills a more basic and practical need. Environmentalism is about change, but the critics neglect to appreciate that these ideologies work on another level as well. Rather than just political rhetoric designed to change societies that "Greens" don't like, environmentalism also provides citizens with interpretive knowledge about how the basic institutions of society work and offers tactics for change. More than just wild-eyed utopian contentions, the ideology of environmentalism helps to orient many citizens to a problematic situation around them—in particular to the question of who they should believe and trust.

Missing is the recognition that in some situations, people need just such an orientation. When confronted with risky circumstances, people look for help in understanding how such circumstances came about, how the system that created them really works—not just how officials say it works—and thus who or what they should worry about. Relying on established ideological perspectives offers quick, shorthand guidance. Much more than half-truths distributed to defend a particular set of interests, such belief systems represent the interpretive synopsis of a long history of experiences with social phenomena. In a complex world, they serve to simplify basic messages down to a few manageable premises that can serve as guidelines for thought and action (which is not to say that people shouldn't or don't reflect on the content of these ideologies, at least over time). In uncertain situations that require action without the luxury of time, they help to give people a basic orientation. And the ideologist, of course, makes a point of being there to help them. This holds as much for the ideologist of the free market as it does for the environmentalist.[6]

If environmental rhetoric is as utopian as Rubin and the critics contend, one also has to ask how the general public comes to be so duped by it, or at least as those critics would have us believe. Few would portray the publics of Western industrial nations as radical or utopian. Indeed, this is just the concern of the critics; the environmental movement, they complain, is out of touch with their publics. So how does the movement manage to gain any influence at all, let alone rise to the position of a powerful voice in an unprecedentedly short period of time?

In answer to the question, two points need to be brought to bear. The first is that the most influential environmental groups, such as the Sierra Club or the Audubon Society, are scarcely radical. Indeed, radical environmentalists like those identified with smaller groups such as Earth First! or the environmental justice movement generally, as we saw in chapter 6, have long accused the mainstream environmental groups of having a far too cozy relationship with industry and business. The second is that these groups tell a story about the relationship of industry to the environment that resonates with the experiences of a sizable number of people in the society. One need not identify a conspiracy against the public to notice the close relationship of environmental degradation to industrial development. Business and the profit motive are clearly tied to the problem. For those who are prone to a skeptical view of business's role, environmentalism helps to flesh out the story. It also serves to reassure such people of their own less articulated view. In this sense, it works something like advertising. It helps the movement hold on to its existing followers as much or more than it makes new converts.

Where, then, does this leave us? Given the limits of science in questions of public policy, coupled with the citizen's reliance on social ideologies, how should we approach deliberation about environmental risk? The critics of environmentalism continue to argue that more and better science is the answer. Recognizing the limits of existing science, Rubin argues that the scientists have a greater responsibility to point out the shortcomings and criticisms of their analyses. But this misses the role of cultural rationality and the problem that it addresses. The solution is to be found not in greater scientific clarification but in answers to questions about the way of life. Whereas the critics take this to mean a call for a different society, significant numbers of people are worried that the society they live in and accept is not working the way its leaders tell them it does. For this reason, the attempt to rule out social ideology can only miss the crucial part of the problem. The challenge ahead is not just more science but rather how to better understand the interactions between science and ideology — facts and values — and most importantly how to integrate them systematically in a more comprehensive analysis.

Conclusion

The call for increased participation involves more than just getting larger numbers of people to come to meetings. It also involves bringing another kind of rationality to bear on the decision-making process. Whereas technical experts have tended to see the involvement of more people as merely clogging up the process, they have failed to recognize that it also brings in the insights of cultural rationality, a consideration that science has failed to appreciate.

In this view, we see that social ideology plays an important, even necessary, role in making informed judgments in the world of action. This leaves us with the question: How can we reformulate our methodologies to integrate the technical and the cultural? The question becomes a central consideration in parts 3 and 4.

Local Knowledge and Participatory Inquiry

Methodological Practices for Political Empowerment

Political action on the side of the oppressed must be pedagogical
action in the authentic sense of the word, and therefore, action
with the oppressed. [The] real humanist can be identified more
by his trust in the people, which engages him in their struggle,
than by a thousand actions in their favor without that trust.
— Paulo Freire

Part III turns to citizens' efforts to engage in research pertaining to
their own issues and interests. The discussion concentrates on methods that have emerged to facilitate cooperative endeavors of citizens
and experts, as well as the nature of the knowledge that citizens bring
to the inquiry process.

Because there can be confusion surrounding the concept of participation, it is important from the outset to be clear about how we
approach citizen participation. Although participation is a political
virtue in and of itself, as a practice it is a challenging and often frustrating endeavor. Collective citizen participation is not something that
can simply happen. It has to be organized, facilitated, and even nurtured. Without concern for the quality of participation, it is better to
forgo the effort. Such endeavors will almost surely fail, and the failure
will only offer the critics of participation ammunition to suggest the
foolishness of the commitment.

Before advocating or entering into a participatory project, we need to recognize that citizen participation schemes rarely follow smooth pathways. Working with local citizens is far from easy. Moreover, local people may themselves be highly skeptical as to whether it is worth investing their time and energy in participation. In some situations, participation can also prove to lack local relevance. Other experiences, especially those in Third World countries, show that community participation can carry more significance for outsiders than it does for the poor in those areas. Within communities, not everyone will be able to participate, nor will everyone be motivated to become involved. Even if there is interest there may be time barriers. And so on.

What is needed is careful research into what citizens can in fact do, what kinds of institutional reforms will help them do that, and in which kinds of policy domains such activities are appropriate. Most of what we know about these issues is based on the conventional wisdom. From this view, significant participation is largely unrealistic. In fact, however, we already have a good deal of evidence to show that citizens can do much more than they are normally credited with. Moreover, failure to participate is often as much a manifestation of institutional processes that either hinder it or render it meaningless. Thus to argue that citizens do not participate does not mean they *can't*. The first challenge, then, is to explore the degree to which the average citizen's capacities to participate might be either facilitated or extended (Webler 1999). Toward this end, we need to problematize the question of participation (Cornwall and Jewkes 1995).

Saying that citizens are capable of participating, however, is different from saying that they should always participate. There is no shortage of evidence to indicate that participation for its own sake can be misguided. Experience shows that citizens will agree that participation is highly useful in some cases and not in others. On the one hand, we discover that participation may help us break through a variety of "wicked" or "intractable" problems — those in which we are not even sure what the problem is, let alone the solution (Fischer 1993). On the other hand, participation can waste a great deal of time and lead nowhere in complex technical issues.

Citizen participation, in short, is a complicated and uncertain business that needs to be carefully thought out in advance. In search of assistance about how to reflect on and organize such participation, we can gain substantial insights from experiences with participatory research and its practices.

Chapter 8 offers two case illustrations of participatory inquiry. The first continues with the problem of NIMBY. Offering a practical illustration of how citizen participation can play an important role in NIMBY-related environmental problem solving, the first case focuses on experiences with a health-oriented participatory methodology, "popular epidemiology." Throughout the story of the struggle between the citizens of Woburn, Massachusetts, and their state government, the chapter shows how citizens and experts can cooperate in an effort to empirically establish the existence of toxic wastes in their local community. The discussion also suggests the ways in which popular epidemiology, an emerging method in environmental struggles, can serve not only as a strategy for researching the local context of health problems but also as a method for critiquing society.

The second case turns to Kerala, India, and describes the process of people's planning and participatory resource mapping that has been introduced there through joint efforts of the state government and the people's science movement. In particular, the case points to the ways in which participatory and scientific planning can systematically be integrated, as well as how participatory local inquiry can be designed to augment the larger political decision structures of a society. Chapter 9 examines more specifically the politics and practices of participatory research methods, of which popular epidemiology is an important variant. Emerging in significant part as a critique of expertise more generally, participatory research is the product of political activists and progressive professionals identified with social movements, in particular Third World movements concerned with issues of environment and technology. As such, the method has developed as an empowerment strategy designed to help less-privileged citizens in their struggles to better understand and confront the realities and choices that shape their own interests and concerns. Toward this end, the discussion explores the role of participatory research as a critical practice, the nature of the citizen-expert relationship basic to its exercise, as well as the methodological strategies and techniques associated with it. In this model, the expert emerges as a facilitator of citizen learning, posing questions and presenting information that assists citizens in their own efforts to examine their interests and to answer their questions in their own terms. The chapter closes with a discussion of the import of such methods for policy analysis more conventionally understood.

In chapter 10, the analysis turns to the nature of the primary product of participatory research, citizens' "local knowledge." Local knowledge is here understood as knowledge about a local context or setting, including empirical knowledge of specific characteristics, circumstances, events, and relationships, as well as the normative understandings of their meaning. As such, it is a type of knowledge that owes its status not to distinctive professional methods but to casual empiricism, thoughtful reflection, and common sense. The discussion traces the role of such knowledge from prescientific times and examines its place in the agricultural practices of contemporary Third World countries. Taking note of the recent upsurge of interest in indigenous local knowledge by a range of scientists and experts involved in agriculture, ecology, and agroecology, the chapter explores the implications of local knowledges for both scientific inquiry and community problem solving. Contrary to the conventional wisdom, local knowledge comes to be seen as a complex, valuable source of largely untapped knowledge that speaks directly to specific kinds of problems. A variety of examples are offered, including the role of local knowledge in the life of potato farmers in the Andes and its contribution to wildlife conservation in South Africa. The chapter closes with a discussion of the broader implications of such ordinary local knowledge for public policy inquiry more generally.

8. Citizens as Local Experts

Popular Epidemiology and Participatory

Resource Mapping

I couldn't believe their reports. I couldn't believe them because
I smelled the stuff and it was vile. Without a Ph.D. in chemistry,
without knowing what was in the water, I knew something was
wrong. The morning that I suspected my water was bad,
I condemned my own well. — New Jersey citizen struggling
against a toxic dump site

Risk assessment, as we saw in the previous chapter, has failed to find a way to circumvent or avoid NIMBY. In fact, by attempting to skirt citizen participation and the democratic process, risk assessment has mainly succeeded in aggravating the conflict. Portraying the method as technocratic and elitist, many environmentalists have instead emphasized more participation. The goal, according to such participatory environmentalists, is to render obsolete the "expert" status of government's and industry's scientists by making "every citizen conversant at all levels of the environmental debate" (Thornton 1991, 15).

In chapter 7 we saw that there is in fact reason to believe that the call for participation is more than just political rhetoric. In cases such as Swan Hills, participation has suggested a way around the kind of political stalemate that has come to define such NIMBY struggles over hazardous facilities. Contrary to most expert opinions, this case provides important evidence suggesting the answer to be more, rather than less, democracy.

At the same time, we saw that citizen participation opens up the decision process not only to a range of new perspectives but to another kind of rationality as well. A more careful look at the dynamics of citizen decision making posed the question of how to integrate sociocultural rationality with the technical perspectives of the experts. This chapter examines these issues in more detail by exploring the ways in which citizens have been able to grapple with the challenge of research in a local context. First, the discussion looks at the abilities of the citizens of Woburn, Massachusetts, to come to grips with an outbreak of leukemia caused by toxic chemicals in the environment. Second, it examines the experiences in people's planning and participatory resource mapping that have taken place as part of a progressive political reform strategy initiated by the government and the people's science movement in Kerala, India.

Participatory Expertise

Many experiences from both social movements and institutionalized deliberative practices show citizens to be much more able to deal with complicated social and technical questions than the conventional wisdom generally assumes (Doble and Richardson 1992). No case, for example, better illustrates such capabilities of citizens than the gay movement's struggle against the spread of AIDS. As gay AIDS activists have shown, citizens can not only learn a great deal about science but also take charge of their own experimentation when deemed necessary. In *Impure Science*, Epstein (1996) documents the degree to which the boundaries between scientific "insiders" and lay "outsiders" have crisscrossed in the struggles to find a cure for AIDS, or what he calls "credibility struggles." In addition to revealing how scientific certainty is constructed or deconstructed, his investigations show nonscientists to have gained enough of a voice in the scientific world to have shaped to a remarkable extent National Health Institute–sponsored research.

Another interesting example, as we saw, is found in Hill's study of efforts to site and operate a nuclear power plant in California. Drawing on extensive surveys and interviews, Hill (1992) shows the degree to which citizens were able to participate in sophisticated policy decisions concerning complex technical issues. He finds that the issues and questions are often posed in languages that are alien to the nonexpert citizen. But once this barrier is overcome, as he argues, comprehending and judging the basic elements of a policy argument about a complex technology are inherently no more complex than what the average citizen does when he or she successfully runs a small business or a family.

Further, although perhaps unintentional, support for Hill's argument can be drawn from a project of Aaron Wildavsky and his former students at Berkeley. No friend of environmentalism, Wildavsky often portrayed the movement's positions on ecological democracy as irrational, even at times fanatical. In an experimental project concerned with citizens' abilities to address rationally the questions of risk, he and his graduate students showed that citizens can learn and use enough science to judge questions of technological risk for themselves (Wildavsky 1997). They concluded that there is no reason to believe that citizens are incapable of mastering the necessary science, at least if they are willing to devote sufficient time and energy to it. Although the effort was designed to counteract what Wildavsky saw as the citizen's reliance on ideology over science, the result of the work differs sharply from the standard technocratic perspective — namely, that citizens are altogether incapable of participating at the technical level of discussion.

Basic to all such cases is the development of cooperative relationships between citizens and experts. Rather than being a matter of citizens merely going it alone, nearly all such cases reveal the involvement of a citizen expert of some sort. From the outset of NIMBY and the citizen movements against the siting of toxic wastes, one almost typically finds present in such struggles a professional expert who assists the community in answering its own questions on its own terms (Levine 1982). Such experts have emerged to help communities grasp the significance of evolving developments, think through strategies, and even directly confront a community's opponents (Edelstein 1988). In the case of Swan Hills, for example, we saw that a local community hired its own experts and set up regular discussions of both the safety and desirability of a hazardous waste incineration plant. In the process, community members balanced safety and health considerations against economic advantages and considered the reliability of the plant management, the necessity of such a facility for the region as a whole, the role of toxic chemicals in modern industry in general, and compensation schemes for the local community, among other issues. In the end, local citizens set out their own criteria for accepting or rejecting the new facility. The result was an agreement to proceed, with the understanding that the community could withdraw the decision any time it felt the operators were in violation of its requirements (Rabe 1994).

Another important experience was the toxic waste crisis at Love Canal in upstate New York — the case that brought national attention to the problem of hazardous waste in the United States. Accounts of the Love Canal Homeowners Association's struggles with state and local officials emphasize the work of a biologist who helped the community association to reinterpret government data, develop the capacity to collect additional information, and interpret this information credibly inside and outside the neighborhood (Paigen 1982).

Such experiences have given rise to a form of community risk assessment designed to bring local residences more directly into the investigatory processes (Chess and Sandman 1989). A direct outcome of the Love Canal experience was the formation of a national organization designed to provide just such alternative expertise to other NIMBY groups across the country. The Citizen's Clearinghouse for Hazardous Wastes was started by Lois Gibbs, the Love Canal homemaker who organized the community and extracted major concessions from the State of New York and the federal government (Gibbs 1982). With only a high school education and no former experience in such matters, Gibbs went on to establish a major Washington-based organization to assist other communities across the country in struggles against toxic wastes.

In recent years, the Citizen's Clearinghouse has interpreted its activities as a contribution to the environmental justice movement. Among its various projects, the clearinghouse offers instruction and advice to local communities for dealing with technical dimensions of the hazardous waste problem, in particular the problem of incineration (Collette 1987). Fundamental to such instruction is training in how to talk to experts, how to understand the expert's research findings, and in some cases how the community can derive its own calculations. But the clearinghouse's function is only to facilitate. Its staff advisers have made a practice of waiting for such groups to find them, rather than directly attempting to organize community groups across the country.[1] Such participatory consultation serves both to broaden citizens' access to the information produced by scientists and to systematize their own "local knowledge." In the United States, the most progressive example of participatory inquiry in environmental research — or perhaps in any policy domain — has taken the form of "popular" or "lay" epidemiology.

Community Risk Assessment in Woburn

Popular Epidemiology

Epidemiology is generally the initial step in a health-related environmental risk assessment. It is defined as "the study of the distribution of a disease or a physiological condition in human populations and of the factors that influence their distribution" (Lillienfeld 1980). The data of such a study are used to explain the etiology of the condition and to provide preventive public health and clinical practices to deal with the condition.

Popular epidemiology, by contrast, is described as "a process in which lay persons gather statistics and other information and also direct and marshal the knowledge and resources of experts in order to understand the epidemiology of disease" (P. Brown 1990, 78). Beyond this important difference — namely, that laypersons collect the statistics — popular epidemiology differs from conventional epidemiology in another significant way. It includes attention to the basic structural features — social and communicative — of both the community and larger society of which it is a part. Popular epidemiology is also explicitly political and activist in nature. In Brown's (1990, 84) words, popular epidemiology is a "highly politicized form of action . . . [that] is also a form of risk communication by lay persons to professional audiences, and as such demonstrates that risk communication is indeed an exercise of political power."

There is no better example of participatory research than that which took shape in Woburn, Massachusetts, in the late 1970s and early

1980s. Although Woburn is scarcely the only experience along these lines, it represents one of the most highly developed illustrations of participatory inquiry to date (P. Brown 1990; P. Brown and Mikkelsen 1990). In response to the discovery of the presence of toxic wastes, coupled with an inordinately high degree of childhood leukemia, community members in Woburn mobilized to investigate the problem and to challenge state and local authorities with the data they were able to assemble.[2]

The residents of Woburn were shocked in 1979 to learn that construction workers had found more than 180 large barrels of waste materials in an abandoned lot alongside a local river. In reaction to citizens' concerns, the Woburn police department notified the State Department of Environmental Quality Engineering, which, after investigation, discovered high levels of carcinogens in several local water wells and ordered them closed. Additional investigation, moreover, revealed that a few years earlier, an engineer from the state had detected high concentrations in the same water supplies, but state officials had failed to investigate the matter. Local residents further learned that the city had received complaints about the water (e.g., a foul taste, dishwater discoloration, and peculiar odors) and had commissioned a consulting firm to examine the matter, which in turn led to a state investigation. At the time, it was thought that the problem stemmed from the interaction of chlorine with other minerals in the water supply. City officials thus ordered a change in the town's chlorination system.

The community's efforts to come to grips with the problem had in fact predated the closing of the wells. Anne Anderson, a local resident whose son had been diagnosed with leukemia, began collecting stories and information about other illnesses through discussions and chance encounters with victims at her son's hospital and in local shopping establishments (P. Brown 1990, 79). Given the surprising number of cases that surfaced in her inquiries, she started to speculate about the origins of the leukemia cases; perhaps they had resulted from something in the water supply. She registered her concern with the state agency but was informed that the agency could not test the water on the basis of citizen requests.

Some six months latter, the Woburn press reported that the state agency had itself discovered another toxic waste site in the area but had again decided to withhold information. At this point, a local minister grew skeptical about the state's earlier reports and became suspicious of its lack of interest in further investigation. Together, he and Anderson placed an advertisement in the local paper; they asked fellow citizens with knowledge about other leukemia cases to contact them. Stunned by the response, they consulted a local physician and proceeded to plot a map that clustered the cases. Convinced of the significance of the clustering, the physician notified the Centers for Disease Control (CDC) of the apparent danger. At the same time, the community activists passed along the findings through the local media and convinced the city council to request that the CDC initiate an investigation (P. Brown and Mikkelsen 1990, 12). Furthermore, Anderson, Young, and about twenty other citizens founded For a Clean Environment (FACE) to mobilize community concern about their findings.

Shortly after the Woburn city council made a formal request to the CDC, the Massachusetts Department of Public Health submitted a report that took sharp issue with the Anderson-Young leukemia map. According to the department, there was no reason to take the map seriously. Said to show no significant evidence of a cluster, the map was dismissed as the work of amateurs. Despite this setback, the community activists were bolstered by a growing national awareness of toxic hazards in the environment, as well as community efforts in other places. In fact, in the context of this growing climate of concern, Anderson and Young were invited by Senator Edward Kennedy to testify at congressional committee hearings pertaining to the toxic waste problem in the country as a whole.

Eventually, in response to the local physician and the city council, the CDC dispatched a scientific team that worked with the Massachusetts Department of Public Health to investigate the Woburn complaints. About six months later, the researchers submitted a report attesting to the fact that cases of both leukemia and kidney cancer in the area were higher than normal. Nonetheless, they concluded the data to be inconclusive. In particular, "the case-control method failed to find characteristics that differentiated victims from nonvictims. Further, a lack of environmental data for earlier periods was an obstacle to linking the disease with the water supply" (P. Brown 1990, 79). But the families and friends of the victims were once again unwilling to accept the conclusions of the report and began to question the scientific study itself. As one journalist put it, a "layperson's epidemiology" began to emerge (DiPerna 1985, 106–8).

The first major step toward a more sophisticated lay investigation came when an interested Harvard professor invited Anderson and Young to discuss their findings in a seminar at the university's School of Public Health. Present was Marvin Zehlen, a biostatistician, who became intrigued with the case. In an effort to elicit more conclusive data, Zehlen and a colleague decided to undertake a more detailed investigation of the health problems in Woburn, in particular environmentally related reproductive disorders and birth defects. To do this, the Harvard biostatisticians and the FACE activists officially agreed to team up in what was to become a major epidemiological study. FACE coordinated some three hundred volunteers to administer a telephone survey designed to reach 70 percent of the population. The Harvard scientists, in turn, supplied the volunteers with training on how to conduct the health survey, in particular how to avoid bias in asking questions and recording answers. In the view of Brown and Mikkelsen (1990), the project became a prototype for a popular epidemiological alliance between citizens and scientists.

Altogether, the scientists and citizens assembled research data that included detailed information on twenty cases of childhood leukemia, a careful examination of the Department of Environmental Quality Engineering's data on the regional distribution of water from the wells, and the results of the community health survey. The biostatisticians, moreover, conducted a variety of analyses to detect bias in the data. At the end of the research process, the team concluded that leukemia was in fact significantly associated with exposure to water from the well.

The public distribution of the Harvard/FACE report immediately encountered harsh criticisms from the CDC, EPA, and the American Cancer Society. Even members of the Harvard Department of Epidemiology took issue with the findings. Many of the criticisms, to be sure, were based on legitimate scientific concerns. In response to the main criticisms, Zehlen and his colleagues pointed out that in such a study there would never be sufficient numbers of each of the numerous defects to fully satisfy statistical procedures (P. Brown 1990, 81). What is more, they showed that their data groupings were appropriately based on the chemical literature concerning birth defects, and they argued that if their groupings were in fact incorrect, they would not have uncovered positive statistical correlations.

The harshest criticisms were directed at the very idea of public participation in science. Because of its "unorthodox methods," the study was said to be biased and thus invalid. The main complaint was that it relied on a health survey conducted by nonscientific citizen volunteers, who in turn were motivated by community interests. Whereas science is said to be impartial, the critics charged that the research was founded on political goals.

For present purposes, however, it is exactly this characteristic that made the case interesting. All things considered, the affected families had confirmed through their own efforts the existence of a leukemia cluster and demonstrated that it was traceable to industrial waste carcinogens that had leached into the drinking water supply. They were able to initiate a series of actions that resulted in a civil lawsuit against two major corporations, one of which the court judged to have negligently dumped its chemical waste products.[3] The legal case moved to a subsequent stage in which the plaintiffs were obliged to prove that the chemical wastes were in fact responsible for the leukemia cases. As this part of the process got under way, the judge determined that the jurors had not adequately comprehended the epidemiological and environmental data crucial to the case and ordered it to be retried. To avoid the possibility of an extremely punitive verdict, the corporation at this point agreed to an out-of-court-settlement with the community plaintiffs. In short, the efforts of FACE paid off.

Not only had the case helped to demonstrate nationally that corporations have the responsibility for toxic wastes and their resultant health effects, but it also offered a valuable example of lay detection and communication of risk to scientific experts and government officials. The exercise has been described as a "prototype" for "low-cost" epidemiology (P. Brown and Mikkelsen 1990; Raloff 1984). For Brown, such efforts are best referred to as "popular epidemiology."

Popular Epidemiology as Participatory Praxis

Popular epidemiology, as a collaborative inquiry involving citizens and experts, is not only an intervention in public health discourse; it is also a method for a critical praxis (Novotny 1994, 1998). As a critical "people's science," it helps to redress losses of public accountability resulting from technocratic uses of dominant forms of scientific and technical discourse.

Emerging in the context of environmental struggles rather than in academe, popular epidemiology takes as its starting point the fact that traditional epidemiology frequently obscures the interrelationships between physiological and sociological factors in the analysis of health disorders. Because traditional epidemiology tends to limit itself to the broad and generalizable trends related to the incidence of health problems, it overlooks the disparate concentrations of such disorders in particular localities. As such, it neglects disproportionate risks in occupation and workplace exposures assumed by low-income and working-class persons of color, especially women.

Popular epidemiology thus challenges the decontextualized individualism of traditional epidemiology by focusing attention on the connections between specific localities — workplaces and communities — where the health of people is endangered. It does this by combining traditional sociodemographic and historical research with community studies that pinpoint health effects of community-based industrial and environmental hazards. The basic strategy of popular epidemiology, in this respect, has been the use and development of the "community health survey." The community health survey is essentially a method designed to help citizens document for themselves the environmental problems in their own neighborhoods (Gibbs 1986). These surveys are citizen-led health studies of the patterns and concentrations of health disorders suspected to be linked with community environmental and workplace hazards. One of the unique aspects of such community health surveys is their ability to construct the environmental health hazards facing communities in social and cultural terms that are comprehensible to the residents. Perhaps the most important aspect of such surveys, however, is their actual empirical impact on the understanding of an epidemiological problem. Such research has the ability to bring to the fore environmental data and circumstances — the facts of the situation — that traditional studies cannot or will not reach. That is, it directly contribute to empirical study of the problem itself.

Because of its closeness to the community, especially politically activated communities, popular epidemiology's ability to draw connections between environmental, occupational, and residential health disorders has made it an effective strategy for political mobilization. By drawing public attention to concentrations or "clusters" of public health disorders, such research can be used to pressure government officials, public health professionals, and private industry to respond to the health concerns of residents. By connecting diffuse community grievances with immediate problems in surroundings familiar to workers' families and friends, popular epidemiology serves as a methodological strategy for the kinds of consciousness-raising that often leads to direct political action.

Even more important than conventional pressure tactics is the transformative and empowering impact popular epidemiology can have on community members. Citizens engaged in community health surveys undertaken in conjunction with community-based political organizing efforts, according to one organizer, learn to better understand the destructive roles that industry and government play in ecological degradation. It can lead to the recognition, as biologist Richard Levins (1990, 117) has put it, that the question of what constitutes a "health issue is resolved not by some scientific method but in struggle." Popular epidemiology, practiced in this way, is also a strategy for political empowerment. For this reason, it has been increasingly conceived as a methodological tool for environmental justice. Its emphasis on the "unspoken categories" of class, gender, and race in environmental and occupational health fits squarely into the environmental justice movement.

At this point, we turn to the second case, a story of people's planning and participatory resource mapping emerging from struggles for economic justice in Kerala, India. Here I further examine not only how people can engage in local participatory research but how such research can be built into the larger political decision-making structures of the state as a whole.

People's Planning in Kerala, India
Participatory Resource Mapping

Located on the southwestern coast of India, Kerala is one of the poorest states in the country. Its densely packed 29 million inhabitants live on a per capita income estimated to be about $300 a year (Kapur 1998). Since the middle to late 1970s, the successive governments of the state have pursued social and redistributive policies that have surprised students of development (Franke 1993). As a result, Kerala's citizens enjoy a level of social development that can be compared favorably with more-developed middle-income countries. Moreover, the state has implemented a system of decentralized development planning that can only serve as a model for others elsewhere.[4]

It is difficult to summarize the complex set of factors that have led to these developments in Kerala. For present purposes, however, the story can be understood in terms of four basic factors that came together in the late 1980s and early 1990s. One is the failure of long-term efforts on the part of the central and state governments of India to make good on a constitutional commitment to decentralized planning (Thakur 1995).[5] A second is a widespread concern about the urgent need to find new ways of dealing more effectively with the pressing and persistent problems resulting from the government's inability to bring economic and social development to the majority of the Indian population. A third has been the interest of a coalition of left-wing parties in Kerala to bring people closer to their local governments. And last but not least there has been a persistent need to find new ways to deal with the problems of development at the local level.

It was against this backdrop that the newly elected communist-led Left Democratic Front (LDF) in Kerala resolved to initiate a People's Planning Campaign to empower the *panchayats* (roughly equivalent to a rural county in the United States) and municipal bodies to draw up the Ninth (five-year) Plan to be submitted to the planners in Delhi. The idea initially emerged as part of a larger debate. With decades of redistribution struggles behind them, efforts that were remarkably successful in flattening out the distribution of income in Kerala, many leaders of these parties felt that few major gains could come from further emphasis on redistribution programs, at least not at that time. This raised the question of how best to proceed on other fronts, such as land reform, environmental protection, public health, and women's equality. Was it possible to channel the energy and resources of the people into direct action for economic and social development? (Franke 1993, 279–80).

Toward this end, the LDF tried a number of experiments to answer these questions. Collectively identified as the New Democratic Initiatives, one of them was the People's Planning Campaign. Kerala's State Planning Board made an unprecedented move; it announced that 35 to 40 percent of the planning activities would be formulated and implemented from below and allocated to the local level an equivalent share of the planning resources.

Drawing on the extensive network of voluntary organizations and mass movements in Kerala, the People's Campaign has sought to motivate and bring together local representatives, officials in the various line departments, governmental and nongovernmental experts relevant to the local planning process, and the mass citizenry. Civic groups and local representatives, many of whom had heretofore been little more than the passive objects of development planning, were mobilized to work to improve the daily lives of the citizens of Kerala. The officials of the government departments, along with relevant professionals, were instructed to decentralize their planning responsibilities and to cooperate in a new democratic project.

To create a political environment conducive to the process, civic organizations were called on to assist in mobilizing their members through publicity strategies and sociocultural "conscientization programs," a concept drawn from the work of the Brazilian educational theorist Paulo Freire. In addition, an autonomous media center with support from the state government initiated a "total communication program" designed to stimulate citizen involvement. Apart from the electronic and press media, the campaign employed a range of other audiovisual cultural approaches based on folk arts. Drawing on techniques developed in an earlier and highly successful campaign for "total literacy," participatory street theater, dances, and local festivals sensitive to the local culture milieu encouraged citizens to take part.

The basic planning task of the People's Campaign has been the local development of an integrated plan. Toward this end, local bodies are directed and assisted in prioritizing and preparing a scheme of integrated programs that constitute the basis for formulating the Ninth Plan for Kerala at the state level. With planning funds and resources from the central agencies, the local bodies are directed to assume the planning responsibilities for themselves. Using both scientific and participatory processes, they identify the needs of the people, assess the development problems facing their areas, survey the local resources available, establish feasibility development plans for priority projects, and integrate them into a local five-year plan.

Empowering Citizens

The People's Plan

This decentralized planning process constitutes a hierarchy of deliberations moving from the village to the block and district levels upward to the State Planning Board. To help ensure maximum participation, the assemblies are held on holidays. Squads of volunteers visit each household to explain the program and urge the people's participation at the assemblies. The goal is to encourage at least one member of each household to attend the meetings.

The actual process commences with the formation of various groups to deal with specific issues, such as agriculture, schools, and environment. Present in each group are trained resource persons who serve as discussion facilitators to guide what is best described as a "semistructured discussion." Information about the local area is gathered by citizen groups, and specific development problems are identified. The citizens are then assisted in analyzing these problems on the basis of their own experience, and to the extent possible, they suggest solutions. The deliberations of each group are summed up at the plenary session of the local convention. The meeting concludes with the selection of representatives to take these local plans and proposals to the deliberations of the "development seminar," which constitutes the next higher stage.

The task of the development seminar is to come up with "integrating solutions" for the various problems identified at the lower-level conventions. In addition to the local citizens' representatives, the key government officials of the areas, as well as invited experts from the locality and outside, attend the seminars. At this level, a report is made, the "Panchayat Development Report," which is based on an analysis of the current development status of the area and a review of the ongoing plans. The plan outlines the development problems in the area and identifies constructive possibilities.

The seminar focuses on general statements of potential solutions to development problems, as it is the job of the third phase of the campaign to convert these solutions into project proposals to be included in the final state plan. For this purpose, task forces of officials and activists again assemble for each of the development sectors. The groups engage in a detailed review of the proposals made at the seminar and draw up working plans, basing them on the necessary technical considerations, time frameworks of the overall process, cost-benefit estimates, and other details mandated by the State Planning Board. For each scheme, the task forces also assess the resources required, as well as the level of available resources. In particular, they examine the extent to which the costs can be met through contributions from local financial and nonfinancial institutions, as well as through the labor of efforts of local volunteers.

The fourth phase of the campaign is the actual formulation of panchayat and municipal plans. Special meetings of the local bodies are convened for this purpose. In the process, experts help in preparing the final document. To facilitate this, the State Planning Board makes available volunteer experts to whom the work groups can turn for assistance. Efforts are made to find solutions that do not depend on state funds. With the help of voluntary contributions of labor, money, and materials, local bodies are encouraged to take up additional schemes of their own.

Continuing in pyramid fashion, the fifth phase of the People's Plan consists of an integration of local plans at the district level. The district plans then constitute the basis for the overall state plan. The state plan is formulated in such a way as to integrate the district plans drawn from the bottom up. The state then allocates its resources to the local levels for the purposes of carrying out the plans.

Of particular importance here is the most basic question that emerged at the outset of the process; namely, where to get the information for the discussions at the local levels. How could planning be turned into a meaningful activity without the requisite information for the process? In most cases, both the quantity and the quality of information has been in short supply. Here the state planners turned to the people's science movement (KSSP) for assistance, an organization that already had extensive experience with the techniques of participatory research. The answer was to be found in process of participatory resource assessment that KSSP developed in conjunction with the Center for Environmental Science Studies.

KSSP: The People's Science Movement

KSSP is a genuinely unique sociocultural movement. Established in 1962, it is the product of a number of scientists and social activists in Kerala who were concerned that scientific information was basically inaccessible to the majority of the people of the state. At the outset, their primary activity involved translating scientific books and other relevant publications from English into the local language of Malayalam and making them available to Keralites, especially books for schoolchildren.[6] The sale of these publications and other books has generated enough income to finance the organization's various other activities. Furthermore, this self-generated income permits the organization to remain free of influences from outside government agencies and NGOs. As a consequence, all of KSSP's members serve as volunteers. Even the president of the organization has a government job by day, devoting his energies to KSSP before and after work.

In 1972 the organization adopted the motto "Science for Social Revolution," and somewhat unexpectedly, it opened the door to what was to become a mass movement with some sixty thousand members. In the process, the emphasis of KSSP shifted from publications to more active efforts to generate a "scientific" questioning attitude in the population as a whole, the underlying goal of which was self-empowerment and change. As a result, interest and involvement jumped.

Two major efforts brought KSSP to national attention. The first was a struggle over the building of a proposed hydroelectric dam in 1984. Drawing attention to the damage the dam would cause to the biodiversity of the state's "Silent Valley" rain forest, activists of KSSP launched a major campaign against the government's efforts to build a dam in the area. Not only did their efforts attract the attention and support of other mainstream members of the scientific community, including international groups, but they caught the eye of the prime minister as well. After a protracted struggle, Indira Gandhi canceled the project with a stroke of her pen.[7]

The second effort involved a literacy campaign. Long concerned with the issue of literacy, in particular scientific literacy, KSSP decided to involve itself in a literacy campaign sponsored by the central government in New Delhi. In 1978, in response to what was seen as a growing literacy crisis in the population, the central government initiated a literacy campaign across the country as a whole. But the effort largely stalled as it bogged down bureaucratically. Little happened, at least until KSSP decided to make the government an offer. Targeting one district in the city of Cochin, KSSP submitted a proposal for initiating a "total literacy" campaign in the district for a minimal price of eighty lakhs rupees, as opposed to the three hundred lakhs rupees that it was estimated to cost the centralized educational bureaucracy for the same activities.[8] Employing the participatory methods that it had developed over the past decade, KSSP in 1989 to 1990 launched a full-scale volunteer effort to bring full literacy to the area.

At the outset, KSSP approached a wide range of local organizations in the district to solicit their support for the project. The response was overwhelmingly positive, with a large number of social organizations offering to supply volunteers for the project. Using participatory educational methods, including those based on the pedagogical theories of Paulo Freire, KSSP's efforts achieved near total literacy in the astonishingly short period of less than three years.[9] So impressive were the results that KSSP was awarded the "Alternative Nobel Prize" in 1996 by the Swedish foundation that has been giving the prize regularly now for some years.[10] Because of the success of the project, coupled with the award, politicians and other organizations have found it difficult to ignore KSSP.

Land Literacy

Participatory Resource Mapping

KSSP's interest in resource use and land mapping began in the late 1980s, when various initial efforts at decentralized community planning were attempted. In an effort to stimulate local planning activities, the government of Kerala had apportioned a small but not insignificant sum of money to the panchayats to spend on their own needs. The experience with the process, however, was less than positive. Local bodies spent their monies on a road or bridge here or there, but in the absence of a clear sense of the needs of the area, let alone a formal development plan, there was no way to relate these projects to an overall set of goals that could help to set in motion the development process. Thus most of the monies were spent in a one-shot manner that didn't lead to anything beyond the immediate expenditure. This led KSSP to begin thinking about how the situation might be rectified.

At about the same time, one of KSSP's members, a local school-teacher, mapped the resources of his own community with the help of local volunteers and the method of participatory rural appraisal (PRA), developed by other participatory research groups for use in the developing world.[11] Intrigued by the process and its outcomes, KSSP members began to wonder if participatory resource mapping might not pose a solution to the problem of development planning in the local areas. But could the process be made more scientific? they asked.

Participatory rural appraisal had largely been developed to offer a quick but reasonably accurate picture of rural resources without getting bogged down in the details of precision. Although PRA has proved to have its uses in the context of underdevelopment, KSSP was interested in infusing the process with more scientific rigor. Toward this end, KSSP approached some of its members at the Center for Earth Science Studies (CESS) in Trivandrum, an institution long engaged in questions of resource management and planning. The question was, Could the local mapping techniques of PRA be combined with the more scientific mapping techniques practiced by the environmental planners of CESS?

Approaching the mapping process as a sociocultural tool for communication among planning experts and local community members, the CESS staff first sought a "base map" from which to begin the process. They identified two candidates for the process: one was a set of typographical maps constructed by an official survey of India; the other, a "cadastral" (or revenue map) designed for tax collections. Insofar as cadastral maps were available for all fifteen hundred villages in Kerala and showed the landholdings in each village, CESS planners adopted it as the basis from which to ask three questions: What are the resources in the areas? Where are they located? And how are they located?[12]

More specifically, these concerns were built around questions pertaining to the features of the land, water resources, and the uses of both. With regard to land, the planners sought information about the specific form of the terrain. For example, did it slope? What type(s) of soil did it have? In terms of water, they focused on characteristics such as a stream or pond, whereas land use raised questions about the type of crops planted and how much land was under cultivation. All were questions for which accurate information was unavailable, and this lack made it difficult to think systematically about development planning.

To these physical questions the CESS staff overlaid the question of infrastructure. That is, what development had already taken place in the area — were there roads, schools, and other community facilities, and if so, where precisely were they located? It also added social data from other sources such as numbers of people in households, earnings, employment, and so forth. Much of the data was obtained or augmented by door-to-door surveys of the villagers.

With the basic design of the planning process in place, CESS planners turned back to KSSP, which in turn took over the assignment of identifying local volunteers for the project, as well as developing and offering a training program on data collection methods. For each village in a particular district, KSSP identified a team of five to eight volunteers. For the selection process, it was decided that educational background need not be a decisive requirement for participation; the questions could be formulated in such a way as to identify local characteristics through a scheme of color coding. The questions, moreover, were presented in local, rather than scientific, terminology.

The training took place in the village itself. Usually it was designed to coincide with a holiday so that the local school could be used for the training session. In general, it lasted three days. On the first day, the KSSP staff explained why the survey was necessary and how the project would be organized. This was followed by more specific instruction by CESS members on the science and techniques of resource mapping and how different types of maps are constructed. On the second day, the volunteers went into the field for on-site instruction in the mapping procedures, in particular how to identify and code specific characteristics of land, water, and resource uses. Through a process described as "learning by doing," the CESS trainers taught the volunteers a number of tests to facilitate the identification process for example, how to roll the soil into a ball with water and to describe what happened in terms of a color chart (Did it get sticky? Did it make a firm ball?). On the final day, the volunteers performed the tasks themselves with the oversight of the CESS trainer in the background. This completed the formal training program.

The volunteers were then sent to the particular village ward to be mapped. Among themselves, they chose a leader, a ward office (usually someone's home), and established a schedule or work plan for the actual mapping, indicating to CESS when they expected to return. As a rule of thumb, the volunteers managed to cover about a square kilometer a day, with the project as a whole lasting four or five days. At the end of the process, CESS planners returned to map on other parameters, less visible to the naked eye of the volunteers, that required expert instruments for measurement—for example, the location of underground water source "potentials."

The result was seven sets of maps. Five of them were constructed by the volunteers; two of them, by the planners. The analytic task was to overlay these maps. Onto the physical land use maps were added relevant survey information about "primary" production and "secondary" social sectors. The resulting map, described as an "Environmental Appraisal Map," was used by the community to work on an "action plan" that served as a component of the larger planning process.

The action plan is thus the product of the local community bodies, assisted by advisory groups made available by the State Planning Board. In the process of examining the environmental appraisal map, in particular comparing its findings and implications with existing practices, the community comes to see how it can change its land use practices. For instance, by learning the locations of underground water sources, they can best determine how to more effectively build and route irrigation systems. Where possible, guidelines for best management practices are produced.

The action plan addresses three questions: What are the problems? What are the future prospects? And what are the gaps between the two?[13] From these questions emerge a sense of what must be done to solve the community's existing problems. In short, there is now a ground for discussion to take place, especially mutual discussions between community members and the planning experts.

Participatory Expertise

State Technical Support

To inform and facilitate the local planning efforts, the state makes available to the panchayats information about all of the ongoing development programs in the state. Even more important, the line department offices of the government prepare a review of their existing development programs in the panchayats, emphasizing the ways they might be coordinated with the communities' own plans.

To facilitate systematic discussions at the development seminars about formulation of the integrated programs, a series of manuals on topics such as watershed management, education and schools, sanitation, drinking water, total energy programs, and environmental protection are prepared and distributed among resource persons. These guidebooks are more than hypothetical exercises. To give the panchayats confidence in the manuals' practicality, they are based on actual local field experiences. Furthermore, local leaders are cautioned that what is needed is not the replication of successful models for other areas but their imaginative adaptation to specific local circumstances.

The state has also organized a cadre of experts to assist the local panchayats in their discussions. All experts are appointed on a volunteer basis and are only advisory to the process. The panchayats may or may not avail themselves of the advice drawn from the full range of sectors — agriculture, education, environment, and so forth. But most have sought such advice out, and it tends to play an important role in the decision process. It is also important to note that an "expert" is defined in Kerala in a broad sense — the term includes not only the civil engineer but also the "wise farmer."

Although the State Planning Board recommends that each panchayat engage in participatory resource mapping, only about 12 percent of the panchayats have completed the mapping process, owing to the time pressures imposed by the requirements of the Ninth Plan. As such, participatory resource mapping has remained as much an ideal as a fully institutionalized practice.[14] It has been encouraged as a parallel activity that will be of use to panchayats in future rounds of planning.

Conclusion

As Woburn and Kerala demonstrate, participatory inquiry is more than a utopian concept. Whereas Swan Hills makes clear the advantages of participatory forums in formulating and implementing solutions to the particularly wicked problems of NIMBY plaguing contemporary environmental politics, Woburn and Kerala more specifically illustrate the ability of a mobilized community to enter the research process. Such participatory strategies speak directly to the concerns of citizen empowerment, democratic theory, and environmental democracy and offer support for both the technocracy critique and postpositivist theory, discussed earlier in the chapters of part 1. With regard to participatory democratic theory, both cases make clear the value of participatory relationships between citizens and scientists. With respect to postpositivist social science, the experiences of FACE underscore the importance of bringing the "local knowledge" of the community to the scientific establishment, as well as the need for scientists to stand in the middle of such processes rather than above them.

Kerala takes this process a step further and shows how the research of the citizens and scientists can systematically be integrated. The process of overlaying the environmental planner's map on that of the citizen volunteers illustrates the way in which both can formally augment and supplement one another. In short, the one mode of inquiry doesn't need to diminish or downplay the other. The Kerala experience also shows that participatory research can be integrated into a larger political decision-making structure. The Kerala State Planning Board has demonstrated how such local efforts, rather than just remaining a local problem-solving strategy of grassroots groups, can meaningfully be connected to higher-level deliberative processes in the formation of the state plan.

Collaborative research relationships are thus more than academic issues; as these cases illustrate, they can bear directly on the outcomes of both policy and research. Indeed, problem solving in the case of "wicked problems" may literally depend on such collaborative methodological innovation. Such methodologies can play an important role in refocusing the ways that lay citizens, scientific experts, and public officials deal with environmental resources and health hazards.

As shown by the experiences presented here, they can assist in mobilizing and empowering communities to identify and communicate resources and risks in ways that have facilitated significant political, economic, and cultural victories.

In the next chapter, I look more systematically at the evolution and development of the methods of participatory inquiry. The discussion sets the stage for an examination of participatory inquiry's role in terms of its specific product, local knowledge, and its contribution to postpositivist methodology more generally.

9. Community Inquiry and Local Knowledge

The Political and Methodological Foundations of Participatory Research

It seems . . . urgent for the planet and for all its creatures that we
discover ways of living in more collaborative relation with each
other and the wider ecology. I see . . . participatory approaches to
inquiry and the worldview they foster as part of this quest.
— Peter Reason

In this chapter, we move from the two cases of participatory inquiry to
a more detailed examination of the methods of participatory inquiry.
The goal is to examine more formally the theory and methods of this
emerging practice, in particular as they relate to local environmental
inquiry. Participatory inquiry, as noted in chapter 2, is a response to
the critique of professional expertise. Basic to the critique is the argument that professional experts have — wittingly or unwittingly —
aligned themselves to elite interests. By and large, professionals have
worked to accommodate others to these interests and views, often
directly at the expense of local citizens or clients. In the name of
democracy and social justice, alternative movements within the professions have sought to develop the practice of "advocacy research."
The result, especially in the environmental sciences, has been the politics of counterexpertise, or political "antidotes," as Beck puts it. Such

a strategy has offered distinct advantages to those without the resources or skills to make known their own interests, knowledge, and views. At the same time, however, experience shows that the practice often failed to foster authentic participation. Even though advocacy brings forth a wider range of knowledge and interests, citizens still mainly sit by and passively take in the exchanges, if they understand them at all. Someone might now be speaking for them, but they largely remain members of the audience. Participatory inquiry, by contrast, has emerged as an effort to bring citizens and their local knowledges directly into the exchange.

The specific focus in this chapter is on the variant of participatory inquiry generally referred to as "participatory research." Participatory research is, in significant part, the product of the work of intellectuals, activists, and progressive professionals identified with Third World communities and the "new social movements" of the more advanced industrial countries (Tandon 1988). As already seen in chapters 3 and 6, such social movements have been the principal agents in the contemporary struggle for participatory democracy. The emergence of ecological and "Green" movements, feminist movements, progressive trade union movements (more typically in Europe and the Third World than in the United States), neighborhood control movements, consumer cooperatives, and worker ownership movements represent an uncompromising call in contemporary society for democratic participation and self-management. In the United States, the most important example of such activities is the Highlander Center in Tennessee, long involved in helping poor communities take charge of their own situation.[1] In more recent years, a small but growing "community-based research" movement employing collaborative research methods has emerged in the United States.[2]

Basic to the efforts of these movements has been the development of an alternative political culture and the participatory institutions and values that sustain it (Offe 1985). As such, they have provided a social form—even laboratory—for experimentation with new sociocultural models, including models of expertise. As alternative movements, they have identified technocratic expertise and its elitist decision-making strategies as primary targets of their countercultural opposition (Fischer 1990). Fundamental to the experiences of these movements have been various forms of experimentation with participatory approaches to science and expertise, especially in the case of the environment.

The classic tensions between expertise and participation are central to these experimental alternatives, as their existence is often a direct response to the impact of new technologies on modern social life (the ecological and antinuclear movements being prime examples). Such experimentation, largely designed to counter the bureaucratic and elitist tendencies that define contemporary political and organizational processes, has in significant part been geared to social movements' emphasis on empowerment and self-help strategies. Emphasizing the development of a nonhierarchical culture, the theorists of these movements — or "movement intellectuals" as described in chapter 5 — have attempted to move beyond the limits of the advocacy orientation by asking a more fundamental question: Is it possible to restructure the largely undemocratic expert-client relationship? Toward this end, one of the key targets of movement intellectuals has been the hierarchical relationship the professions maintain with their clients (Touraine 1965, 1981). Their direct confrontation with this problem offers interesting and suggestive ideas as to how expertise might be adapted to accommodate democratic organizational practices. For this reason, such alternative cultural movements have stimulated a nascent but insightful discussion of alternative expert practices. It is to the literature generated by this discussion that we turn for guidance in an effort to rethink the expert's function in the context of a genuine commitment to participation.

Political and Methodological Foundations

It is difficult today to identify or emphasize one particular approach to participatory inquiry (Reason 1994; Eldon and Chrisholm 1993). The approaches tend to range from action research on the more conventional end of the spectrum to a politically oriented participatory research at the other end.[3] Where action research has largely focused on passively explicating the implicit theories of actors and decision makers and examining their implications for action strategies, especially in managerial settings, participatory research has sought to serve as an enlightenment strategy for raising the consciousness of citizens with common interests and concerns. As such, it has emphasized the political dimensions of knowledge production and the role of knowledge as an instrument of power and control. While a range of approaches fall between these two perspectives, the discussion here will primarily focus on participatory research, as it is the most directly concerned with the politics of the citizen-expert relationship.

The practice of participatory research became particularly prominent in the Third World in the 1970s. This occurred in large part with the recognition that conventional economic and agricultural projects were failing to eliminate or reduce poverty and inequality, the experience with participatory resource mapping in Kerala being a case in point. In response to these failures, as Cancian and Armstead (1992, 1,427) write, "Researchers began to develop alternative approaches that increased the participation of the poor in development programs and aimed at empowering poor rural and urban communities as well as improving the standard of living." Most of these projects have involved farmers and peasants cooperatively working with social scientists and agriculturalists to establish productive and appropriate farming techniques (Rahman 1991; Gerber 1992) (see appendix C). Beyond agricultural projects, Kassam and Mustafa (1982) report on a project in Tanzania, where villagers and participatory researchers studied traditional dance and music to develop small-scale cooperative industries that produced drums and other instruments for sale in urban areas. Patel (1988) describes the organization of a participatory census survey by Bombay slum dwellers to identify greater numbers of residents denied census-dependent services by an official population count. Comstock and Fox (1993) analyze the planning and design of a new town in the state of Washington after a long struggle with the Army Corps of Engineers (see appendix D).

Given the disparate character of a dispersed Third World literature, coupled with the fact that social movements typically stand outside the mainstream of industrial society, it is difficult to present a comprehensive, fully developed model of participatory research, although progress is being made in this direction (Chambers 1997). The task is further complicated because much of the work is conducted outside of universities and seldom, at least until recently, has been published in mainstream academic journals.[4]

It is interesting to note, however, that in more recent years, "participatory research" has become an entry in the *Encyclopedia of Sociology*. There, Cancian and Armstead (1992, 1427), acknowledging its Third World roots, define it as an effort to integrate "scientific investigation with education and political action." Experts and researchers work cooperatively "with members of a community to understand and resolve community problems, to empower community members, and to democratize research." For social researchers who challenge "the traditional values of being detached and value-free and who seek an approach that is less hierarchical and that serves the interests of those with little power," as Cancian and Armstead put it, "participatory research is a valuable alternative."

Participatory research takes its methodological foundations from a variety of sources. Its most important methodological influences include the collaborative methodology of action research, especially its emphasis on social learning; trends in applied anthropological research, including in-depth interviews; ethnography; and participant observation, all of which rely on empathetic interpretation of everyday experience and local knowledge. Theoretically, participatory research draws on work in phenomenological sociology, critical theory, and the writings of Paulo Freire, who is an especially important influence in Third World developments. Moreover, one of the richest contributions to participatory research is in the literature of the alternative educational movement, particularly adult education.

In more recent years, a form of "participatory action research" has gained adherents in sociology. Following the lead of William Whyte (1989), the methodology of this approach closely resembles that of the kind of participatory research under discussion here. In most ways, the main difference is ideological. Those who use the method for progressive or radical causes of empowerment largely adopt the phrase "participatory research."[5] These researchers have tended to focus on consciousness-raising and education, political action, and social change (Uphoff 1992). Those operating more or less within the boundaries of the professional social science community mainly speak of "participatory action research," in part to maintain a link to the earlier action research of Kurt Lewin (Argyris et al. 1985). Whyte (1989), for example, works with managers and workers to find new ways to deal with organizational problems such as cost cutting and redesigning employee training programs. The action components of his investigations are undertaken in cooperation with company management and do little to directly challenge existing organizational power structures through worker empowerment, education, and consciousness-raising. Others might focus on small-scale improvements such as establishing a collective system to reduce water pollution or the misuse of environmentally hazardous pesticides but would devote no time to mobilizing peasant farmers to challenge the ruling elites.

Cutting across the various approaches is a common epistemological orientation. Human beings are cocreators of "their own reality through participation: through their experience, their imagination and intuition, and their thinking and action" (Reason 1994, 324). At the heart of participatory inquiry's critique of conventional scientific methods is "the idea that its methods are neither adequate nor appropriate for the study of *persons*, for persons are to some significant degree self-determining" (Reason 1994, 325). By excluding its human subjects from the thinking that goes into developing, designing, administering, and drawing inferences from the findings, conventional social inquiry alienates itself "from the inquiry process and from the knowledge that is its outcome, and thus invalidates any claim the methods have to be a science of persons" (Reason 1994, 325). Participatory research, by contrast, is fundamentally grounded in the idea that people can help choose how they live their lives. It is, as such, an inherently democratic practice.

Because participatory research can in important respects be defined as a radicalized conceptualization of action research, I turn to a more general examination of the evolution of the collaborative orientation of research. Before doing that, however, I offer a word of caution. The following discussion is largely limited to the theoretical and methodological foundations of participatory research. It is not a discussion about how to do participatory research. Moreover, it is difficult in the context of a theoretical discussion to adequately capture the dynamic of a participatory methodology. Much of the significance of such a methodology derives from the natures and qualities of the interpersonal exchanges it promotes and doubtless can be fully conveyed only in the process of actually carrying out such research.

The Collaborative Orientation

From Action to Participatory Research

The history of the effort to construct a method for participatory inquiry is in part traceable to the action research methodology pioneered by Lewin in the 1940s. Initially developed as a full-scale effort to facilitate a democratic practitioner-client relationship, Lewin's work was motivated by his desire to fashion a mode of inquiry capable of dealing with the social problems of the postwar period, particularly the problem of fascism. Toward this end, he worked out a collaborative research methodology designed to democratize authoritarian decision cultures (Marrow 1969).

Collaborative research, as it emerged from action research, is a "client-centered" methodology designed to facilitate social learning (Greenwood and Levin 1998; Argyris et al. 1985). Formally, it can be defined as a deliberative process in which a practitioner(s) and a client system are brought together to solve a problem or to plan a course of action through the processes of collective learning. Such research proceeds through task-oriented groups, typically involving fewer than a dozen participants. Whereas in the orthodox scientific approach "the problem to be studied is identified by the researcher and, quite frequently, framed in such a way as to take advantage of data already assembled in a library, various agency documents, or a computer," collaborative research takes place in the clients' "natural" setting, drawing on their opinions, judgments, and resources (Sherwood 1978). Essential to the relationship are the following conditions: (1) a joint effort growing out of an interaction between practitioners and clients that involves mutual determination of goals, (2) a "spirit of inquiry" based on publicly shared data, (3) equal opportunity for each party to influence the other, and (4) freedom on the part of both practitioners and clients to discontinue their relationship after mutual consultations (Bennis 1966).

In methodological terms, collaborative action research is a "messy," multimethodological approach that both overlaps with, and diverges from, standard scientific research. Like the scientific tradition, it seeks knowledge that can be empirically generalizable at the same time that it is relevant to specific real-life contexts. Some versions, in this respect, bear a close resemblance to "grounded theory" (Glasser 1992). Like applied research generally, collaborative research demands that knowledge be useful. But unlike both basic and applied research, the collaborative orientation requires that the inquiry process speak to the forming of goals and purposes.

Collaborative research's emphasis on social learning grapples with two of the most sophisticated epistemological issues facing traditional policy science, namely, the relationships of theory to practice and empirical to normative analysis. As a form of knowing intrinsically related to human activity, effective social learning comes from confrontations with social experience. Social learning research thus focuses as much on the social-psychological situation and on the sociocultural contexts of learning as on cognition. Examining problems from the perspective of those engaged in practice, it takes the social environment and the actors' "ordinary knowledge" to be a primary empirical focus in the analysis of learning situations. Relying on the mediating role of small groups, it stresses the crucial function of dialogue in forming collective goals and purposes.

This commitment to connecting theory and practice through collaborative social learning has long been a fundamental tenet of critical theorizing (Friedmann 1987). However, collaborative research, at least as practiced in action research, has failed to fulfill this critical function. Collaborative techniques have in this context been mainly adapted for use in the bureaucratic context of managerial and organizational research. In fact, collaborative research is now a technique and ideology advanced in significant part by management consultants. In the texts on the subject today, one can scarcely find mention of the word "democracy." Instead, practitioners speak of "participatory management" and tout its use as a technique for making bureaucratic organizations more responsive to change. In short, the more inclusive objectives of democratization have disappeared.

If something like the collaborative orientation is required to carry out critical social science research, that something would appear to be "participatory research." Emerging in large part with the new social movements and other citizen initiatives, participatory research's departure from the earlier models of collaborative research is found more in its purposes than in the methodology itself. In sharp contrast to the managerial orientation, participatory research attempts to extend its methodologies to a democratically progressive political orientation. Where action research's collaborative orientation largely developed to assist bureaucratic clienteles, participatory research has evolved from efforts to give a voice to poor and oppressed peoples struggling to improve their lives.

Participatory Research as Critical Praxis
Distributing the Means of Thinking

Basic to participatory research, as with action research, has been the link between knowledge and power. Participatory research, however, has been highly influenced by radical challenges to positivist social science, especially those of feminists (Harding 1986), Marxists (Gramsci 1971), and critical theorists (Habermas 1970b, 1973), among others. Such writers, as we have already seen, argue that the positivists' "emphasis on objectivity, detachment, and value-free inquiry often masked a hidden conservative political agenda, and encouraged research that justified domination by experts and elites and devalued oppressed peoples" (Cancian and Armstead 1992, 1429). Participatory research, as paradigm and methodology, has emerged as a way to "integrate research and theory with political action," as an approach for giving "the people being studied more power over the research." Or as Rahman (1993, 46) put it, as a way of "distributing the means of thinking."

Like action research, participatory research is epistemologically grounded in a phenomenological perspective. Based in experiential knowing, the cooperative inquiry experience "involves a fundamental phenomenological discrimination of persons in relation to their world" (Heron 1981, 158). It seeks, as such, to understand individuals and their problems within their own sociocultural context and the particular "logic of the situation" to which it gives rise. But participatory research seeks to do more. Beyond analyzing the sociocultural logic of action, it seeks to link the experiential situation to the larger social structure. It is an effort, in short, to interpret the situation in terms of the more fundamental structures of social domination that shape it. As such, participatory research casts its findings in the framework of a larger social critique, an epistemological step that links it to critical theory and an "emancipatory interest."

Especially important from the phenomenological perspective is an emphasis on the actor's own "common sense" or "ordinary knowledge" (Lindblom and Cohen 1979). Collaborative researchers draw a distinction between the formal (abstract) knowledge developed in professional inquiry and the actor's informal, contextual, local knowledge, often organized in narrative form and told as stories (Krieger 1981; T. Kaplan 1993). Experiential knowledge, as Heron (1992) explains, is typically ordered into patterns expressed in stories and images. Permitting the integration of a broader selection of social meanings, norms, and values into the analytical process, discourses and narratives enrich the standard quantitative analysis of efficient means to given ends with a qualitative discussion of the ends themselves. Some accept both as valid types of knowledge but recognize each to be geared to different problems or purposes.[6] The task of the researcher is to bring these two types of knowledge together in a mutually beneficial, problem-oriented dialogue. Through dialogue, as Reason (1994, 328) put it, "the subject-object relationship of traditional science gives way to a subject-subject one," in which formal academic knowledge works in dialectical tension with the popular knowledge of ordinary citizens to produce a deeper contextual understanding of the situation.

Participatory research's emphasis on ordinary-language dialogue and storytelling links up with the emerging turn to discourse and argumentation in the social and policy sciences. This "argumentative turn" is itself largely a response to the political and epistemological limitations of policy science's technical orientation, particularly their narrow treatment of normative assumptions and values (Fischer and Forester 1993). In the next section, I examine more specifically the ways in which these interrelated emphases on the actor's social context, the processes of discourse, and group learning are combined in the methodology of participatory research.

Participatory Research as Methodology

In participatory research, attitudes and behavior are more important than methods (Chambers 1997, 212). This, however, is not to say that methods play no role. Indeed, on paper, the basic methodological steps of a participatory research project don't look that much different from those of a standard empirical research methodology. Eldon (1981, 257–58), for example, specifies four critical decisions confronting the participatory researcher: (1) problem definition; what is the research problem? (2) choice of methods; which methodologies will best provide the required data? (3) data analysis; how are the data to be interpreted? and (4) use of findings; how can the outcomes be used? Who learns what from the research findings? "Research is participatory," Eldon explains (1981, 257–58), "when the participants directly affected by it influence each of these four decisions and help to carry them out."

Because participatory research is an inquiry conducted in everyday life, a standard methodological description does not capture the essential experiential side of the practice. While the research process always involves such elements, efforts to codify emergent processes of collaboration and dialogue are destined to fail. More than just an effort to come up with research findings, the method is geared to fostering individual and community empowerment, motivation, and solidarity. De Roux (1991, 44) has effectively captured these two interwoven dimensions: at an intellectual level, the practice must "be capable of releasing people's pent-up knowledge, and in doing so liberate their hitherto stifled thoughts and voices, stimulating creativity and developing analytical and critical capacities"; while at an emotional level, it must "be capable of releasing feelings, of tearing down the participants' internal walls in order to free up energy for action." It is, in this sense, a kind of "consciousness in the midst of action" concerned with "primary" data encountered "on-line" and "in the midst of perception and action" and only secondarily with recorded data (Torbert 1991, 221).

Participatory research theorists and practitioners define the validity of collaborative inquiry in terms of its encounter with concrete experience.[7] Referring to what they call a "critical subjectivity," Reason and

Rowan (1981) describe the validity of this research encounter with experience as resting "on high-quality, critical, self-aware, discriminating, and informed judgments of the co-researchers." Reason (1994, 326–27) has outlined the methodological steps of such inquiry as involving four phases of action and reflection.

Phase 1. Co-researchers agree on an area of inquiry and identify some initial research propositions. They may choose to explore some aspect of their experience, agree to try out in practice some particular skills, or seek to change some aspect of their world. They also agree to some set of procedures by which they will observe and record their own and each other's experience. . . .

Phase 2. The group then applies these ideas and procedures in their everyday life and work: They initiate the agreed actions and observe and record the outcomes of their own and each other's behavior. At this stage, they need to be particularly alert for the subtleties and nuances of experience, and to ways in which their original ideas do and do not accord with experience.

Phase 3. The co-researchers will in all probability become fully immersed in this activity and experience. . . . It is here that the co-researchers, fully engaged with their experience, may develop an openness to what is going on for them and their environment that allows them to bracket off their prior beliefs and preconditions and so see their experience in a new way.

Phase 4. After an appropriate period engaged in Phases 2 and 3, the co-researchers return to consider their original research propositions and hypotheses in the light of experience, modifying, reformulating, and rejecting them, adopting new hypotheses, and so on. They may also amend and develop their research procedures more fully to record their experience. . . .

Compared to other types of researchers, the participatory researcher "is more dependent on those from whom the data come, has less control over the research process, and has more pressure to work from other people's definitions of the situation" (Eldon 1981). As Maguire (1987) points out, this means that the researcher must at the outset carefully consider which segment of the community will participate. As Cancian and Armstead (1992, 1430) explain, the researcher has to get beyond vague generalizations about "the poor" or "the oppressed" and recognize that most communities, poor as well as rich, are "complex and internally stratified." It also means that the researcher has to confront the fact that some people are at times difficult to include.

Furthermore, participatory researchers have to be sensitive to the power differential between the researcher and the researched. As a group, they have more time, money, and specific skills for obtaining information and facilitating group interactions. If community members are to identify and discuss community problems as coparticipants in designing a research project, they need instruction in the methods employed. While this need not require the skill and competence of an expert, a certain basic level of understanding is essential. Without it, such participation will be fairly limited, even superficial; citizens will serve as little more than the researchers' assistants.

These considerations put unique role demands on the professional, ranging from theoretician and expert to colleague and co-producer of knowledge. In each case, the basic determinant of the expert's role choices must be his or her usefulness in facilitating collaborative learning processes. The basic question is this: How can the expert's role facilitate the development of a learning process that, once set in motion, can proceed on its own?

Participatory Expertise

The Facilitation of Learning

The facilitation of participant learning is designed to enlarge the citizen clients' abilities to pose the problems and questions that interest and concern them and to help connect them to the kinds of information and resources needed to help them find answers. Brookfield (1986, 3) defines facilitation as the process of "challenging learners with alternative ways of interpreting their experience" and presenting them with "ideas and behaviors that cause them to examine critically their values, ways of acting, and the assumptions by which they live" (3). Teachers and students, experts and clients, "bring to the encounter experiences, attitudinal sets, and alternative ways of looking at their personal, professional, political, and recreational worlds, along with a multitude of differing purposes, orientations, and expectations" (3). The medium of this interaction is a highly complex dialogue "in which the personalities of the individuals involved, the contextual setting for the educational transaction, and the prevailing political climate affect the nature and form of learning" (3). The dialogue is likened to a "transactional drama" in which the philosophies, personalities, and priorities of the "chief players interact continuously to influence the nature, direction, and form of the subsequent learning" (3). Sometimes this dialogue can even take the form of a story or a drama, using video equipment, drawings, mapping procedures, and various forms of role playing. In Third World countries, especially where literary rates are low, participatory researchers have developed mapping procedures that permit community members to portray and appraise their own information and knowledge about the history, experiences, and conditions of their region.

Hirschhorn's discussion of the professional practices of the alternative human services movement helps to clarify the expert's role as facilitator. Taking up the "crisis of the professions," Hirschhorn focuses directly on the social, emotional, and intellectual distance that separates the professional from the client's experiential lifeworld (Hirschhorn 1979; Rappaport et al. 1985). Indeed, distance has become the source of strident disagreements over the definition of the client's social situation, as well as over who should have the responsibility for determining the issue. Such struggles invariably raise the question of social control, typically leading to acrimonious polemics about the professional's role in the delivery of services. For Hirschhorn, the solution lies in redesigning the professional-client relationship. In keeping with Brookfield's model, the expert must be remade into a facilitator of client learning.

As a facilitator, the expert's task is to assist clients in their own efforts to examine their own interests and to plan appropriate courses of action.[8] In a human services setting, for example, this means the professionals must be skilled in such processes as "role definition, life-course planning, and the collective definition of mutual responsibilities" (Hirschhorn 1979, 187). The assignment is to learn "the necessary and sufficient *conditions* for client learning" and to design and enable "the environment within which clients develop their own conceptions of satisfactory roles" (Hirschhorn, 187). In short, "professionals must become experts in how clients learn, clarify, and decide" (187). Emphasis is thus "on establishing the institutional conditions within which clients can draw on their own individual and collective agencies to solve their problems" (188). The "professional acts as a programmer, mobilizer of resources, and consultant to a self-exploration and learning process on the part of group members" (188).

Essential to the facilitation of empowerment, then, is the creation of institutional and intellectual conditions that help people pose questions in their own ordinary (or everyday) languages and decide the issues important to them. Theorists interested in developing these concepts have most typically turned to models of social learning and discourse. The central focus of such models is how to innovate "inquiring systems" that assist learners in the "problematization" and exploration of their own concerns and interests.

Although participatory research emphasizes "critical consciousness" and "structural change," some like Tandon (1988) warn against expecting to achieve radical social change; "social transformation requires . . . organizing, mobilizing [and] struggle" as well as knowledge. "These researchers point to the values of small collective actions in educating people about local power structures, creating greater solidarity and feeling of power and providing new knowledge about how power is maintained and challenged" (Cancian and Armstead 1992). Some projects are only geared to changing the behavior of individual participants, facilitating critical knowledge, or creating a community network that strengthens the capacity for action.

Facilitation as Problem Posing

In the Third World, participatory research, as we saw in the case of Kerala, has been closely associated with the work of Paulo Freire. Freire's work on "problematization" or "problem posing" is basic to much of the writings on participatory research (Freire 1970, 1973, 66). Problematizing for Freire is the direct antithesis of technocratic problem solving. In the technocratic approach, the expert establishes some distance from reality, analyzes it into component parts, devises means for resolving difficulties in the most efficient way, and then dictates the strategy or policy. Such problem solving, as Freire makes clear, distorts the totality of human experience by reducing it to dimensions that are amenable to treatment as mere difficulties to be solved. To "problematize," on the other hand, is to help people codify into symbols an integrated picture or story of reality that, in the course of its development, can generate a critical consciousness capable of empowering them to alter their relations to both the physical and the social worlds.

Problem posing presents a fundamental challenge to both the traditional teacher-student and professional-client relationship. As Freire (1970, 67) puts it, in the context of critical dialogue, "the teacher-of-the-students and the students-of-the-teacher cease to exist and a new term emerges: teacher-student with students-teachers" (67). No longer is the teacher "merely the-one-who-teaches, but also one who is himself taught in dialogue with the students, who in turn are learning to teach" (67). As the co-producers of knowledge, they become "jointly responsible for a process in which all grow" (67).

In the mainstream literature on professional expertise, the writings of Donald Schon (1983) come the closest to taking up the issue of problem posing and its implications for the professional-client relationship. Schon is fundamentally concerned about the lack of an open and authentic expert-client interaction in policy science. For him, such interaction is key to the reconstruction of expert practices. Like Freire, Schon attributes the failure of professional policy expertise to its outdated adherence to the technical model of rationality and the superior-subordinate expert-client relationship that it requires. Giving rise to one-dimensional, distorted communications between practitioners and their clients, the relationship impedes the activity most critical to effective practices, what Schon refers to as "problem setting." The term is used to connote essentially the same intellectual task conceptualized by Freire as "problem posing."

Problem setting is nontechnical in nature and contrasts sharply with problem solving. Whereas the latter involves technical knowledge and skills, such as those typically associated with policy science methodologies (cost-benefit analysis, systems analysis, program evaluation, etc.), problem setting is fundamentally normative and qualitative. In technical analysis, values and goals are taken as given; in problem setting, analysis focuses on their identification and discovery. Indeed, at times it involves the consensual shaping of new value orientations. An inherently creative exercise, problem setting can be neither explained nor taught from the technical (positivistic) perspective that informs much of professional practice. In this sense, problem setting is better understood as an art form than a science.

In more specific terms, problem setting concerns two interrelated tasks: the determination of the relevant problem situations to be addressed and the theoretical normative "frames" that structure and shape our basic understandings of (and discourses about) particular policy issues, including the criteria appropriate for their evaluation (Schon and Rein 1994). Analytically preceding technical problem solving, problem setting requires professionals to initiate what Schon (1983) calls a "conversation with the situation." Focusing in particular on naming situations and defining the problems that arise in them, "reflection in action" necessitates a new epistemological orientation. The quantitative modes of reason that have shaped policy inquiry must, in short, make room for interpretive modes of qualitative reason.

Participatory Research and Policy Analysis

Although participatory research emerged with local issues such as farming and the use of alternative technologies, in recent years it has begun to play a role in policy analysis more formally understood. An important example is the World Bank's Participation Program. Having learned the relevance of local involvement and participation from many of its Third World investment failures, the bank has in the 1990s taken an interest in the advantages offered by direct local contact with the communities it seeks to assist (World Bank 1994, 1995). Not only are senior bank staff members directed to get to know a particular region better through a week of total immersion in one of its villages or slums, the bank has pioneered a technique called participatory policy assessment, designed "to enable the poor people to express their realities themselves" (Chambers 1997, xvi). Adapting their approach from other participatory research techniques, especially the method of participatory rural appraisal, the bank has now been involved in participatory poverty assessments in more than thirty countries around the world, in particular in Africa (Norton and Stephens 1995).

Participatory poverty assessment represents an attempt to strengthen the bank's analysis of the connections between its assistance strategies and the borrower countries' own programs to reduce poverty. In programs specially designed to inform its policy dialogues with these governments, the bank has sought ways to scale up participatory approaches from the project level to the country level. Toward this end, it has encouraged its operational managers to supplement their conventional poverty research with participatory poverty assessments. Such assessments have not been conducted as discrete research processes but rather have been designed to produce results "that can help to complement, inform and validate conclusions drawn from other kinds of more traditional Bank analysis" (Norton and Stephens 1995, 5). Typically, these discussions among the bank's analysts have focused on "how to best integrate participatory and conventional methods, distinguished as 'qualitative' and 'quantitative' respectively in Bank discourse."

Such participatory assessments have by no means been limited to the World Bank. Indeed, the bank has gotten many of its ideas from nongovernmental organizations (NGOs) and other development institutions. NGOs have designed and conducted a growing number of participatory policy analysis projects — for example, irrigation policy studies in India, wetland management policy investigation in Pakistan, food grain studies in Nepal, forestry issues in Scotland, educational policy matters in Gambia, the relationship between poverty and violence in Jamaica, and land tenure concerns in Madagascar, to name just a few.

These efforts have been judged to offer timely and useful policy experiences, especially when policy decision makers are highly committed, the inquiry is of high quality, and the results are tested against other sources. Offering a voice to the poor, such policy debates become grounded in local realities and citizen interpretations rather than would-be "objective realities" designed by analysts sitting behind desks. Such efforts offer an alternative mode of evaluation that not only provides local information but has proven capable of uncovering insightful, often counterintuitive surprises.

Participatory policy analysis has also emerged in several government agencies in the United States. The most important example is that offered by Dan Durning (1993). Durning has observed and reported on a "stakeholder" approach to participatory policy analysis in the Georgia Division of Rehabilitation Service. In this case, the service assembled a team of the agency's employees to analyze its policy for selecting service recipients and to present advice to the agency's executive committee. In a careful analysis of the process and its outcomes, Durning (1993, 317) concludes that participatory policy analysis is a method that is "well suited for addressing some messy or ill-structured policy issues" (317) (see appendix E).

Participatory Training and Qualitative Inquiry

Participatory research, as well as participatory inquiry in general, poses sophisticated challenges for professional training. This is especially the case as it raises issues of professional conduct — in particular, behavior and attitudes toward client groups — more than it does matters of research methodology. One major issue concerns the role of leadership. Because it is egalitarian — even radically egalitarian — participatory research places unique demands on those who seek to initiate it. As Rahman (1991, 20) argues, insofar as "movements for social change are normally led by intellectuals who are in a position to provide leadership not because of any particular aptitude but because they are privileged by their economic and social status," there are "many dangers of relying on an elite leadership for social transformation: the dangers of inflated egos, the fragility of the commitment in the face of attractive temptations; the problems of the growth in size of the elite class as a movement grows and the danger of attracting new adherents holding altogether different commitments; and finally, the self-perpetuating character of the institutions created to provide leadership."

These changes underscore a central tension in participatory inquiry. Unless someone with the skills, commitment, and time is willing to initiate such a research project, it will almost never come into existence. Invariably such persons are members of a privileged, educated group with elite status in the society, especially so in the developing world. For such research to work, however, it also has to be conducted by people with high levels of personal self-development. One of the most important but often overlooked psychological dimensions is the ability to find "ways of sidestepping one's own and others' defensive responses to the painful process of self-reflection" (Reason 1994, 332). The training to reach such interpersonal skills, or what Torbert (1976) describes as "transformational leadership," should be rigorous and formidable. In response to the challenge, numerous researchers have developed and established training programs to teach such interpersonal participatory inquiry.[9]

Basic to such training must be the tension between participation and the practical demands of competence and leadership. The tension can be understood as a "living paradox" that "we have to live with," for which we have "to find creative resolution moment to moment" (Reason 1994, 335). We can work it out only through an "emergent process that participants are first led through," which they can then "amend and develop in the light of their experience, and finally embrace as their own" (Reason 1994, 335). Heron (1989) sees the management of this tension to involve a never-ending balance among hierarchical structure and the legitimate exercise of authority; group recognition of peers and shared power; and a respect for each group member's right to exercise his or her own judgment.

Once we recognize that participatory research is as much a creative art as a science, we venture into complicated pedagogical territories. Scientific methodology texts, perhaps unfortunately, can be organized like cookbooks. An art form is a different matter. Not only are there no set formulas, but little is known about the creative impulse itself. How, for example, do we educate an analyst to appreciate the range of human folly or the boundaries of human virtue? How do we train the investigator to intuitively sense openings and opportunities in human affairs? If there is an answer, it no doubt includes greater exposure to the creative arts, novels, poetry, culture, and so on. But these are only generalities; the question remains open.

Beyond the creative dimension, however, one requirement is relatively straightforward. In more immediate terms, the professional-client collaboration requires the expert to have special knowledge of the client's needs, interests, and values. Toward this end, there have been numerous projects designed to resocialize professionals to the client's "natural setting." Gottlieb and Farquharson (1985), for example, have spelled out the elements of a pedagogical strategy designed to accommodate the student-practitioner to the ways that clients or citizens deal with their own social welfare and health needs. Most important is the need to eliminate the professional's commitment to, and trust in, the superiority of technical solutions, compelling a reconsideration of professional practices that relegate nonprofessionals to an insignificant or subservient role of patient or client. Specifically, professionals must gain firsthand knowledge of encounters with self-help groups and other collective projects. They must become acquainted with local groups who can motivate and enable others to take it on themselves to cause change. For Gottlieb and Farquharson, they must learn firsthand the empowering effects of mutual aid and assistance.

In curricular terms, then, alternative professional training means offering educational experiences that bring professionals into closer contact with clients' everyday experiences, language, and culture. Such experiences must be designed to wean professionals away from their faith in technique, their adherence to hierarchy, and their reliance on the ideologies of expertise.

The training must also pay special attention to the political dimensions of the facilitator's role in the expert-client relationship. Facilitation and its problem-posing orientation are founded on the long-established but largely ignored assumption that teaching and learning—particularly the creation and change of beliefs, values, behaviors, and social relationships—are acts that give expression to and shape our common humanity (Brookfield 1986). Although clearly political in its import, a commitment to such dialogue is not in and of itself to be confused with a commitment to a specific doctrine or ideology. For participatory researchers firmly committed to democratic values, educational facilitation and political proselytizing are geared to fundamentally different objectives. Political ideologists, accepting their beliefs as the one true way of thinking about the world, proselytize with a predetermined definition of the successful outcome; they simply dismiss diverging views as wrong thinking, bad faith, or false consciousness. By contrast, the facilitator may passionately advance ideas about how people should learn and act but must present such ideas to learners for the same kind of critical scrutiny to which the educator has subjected other views of which he or she is personally critical. The end of the encounter, in other words, is not the acceptance by participants of the facilitator's preordained values and beliefs. Rather, it is to pose problems and questions for critical dialogue and group consensus formation.

Conclusion

The foregoing discussion, of course, is intended as only a sketch of an alternative practice. It does, however, establish a number of basic contributions of participatory research and makes clear its potential as a foundation for the reconstruction of a nontechnocratic alternative. As such, it constitutes a direct political and epistemological challenge to mainstream policy inquiry. Politically, participatory research's dedication to democratic practices provides a dramatic departure from the mainstream commitment to the corporate-bureaucratic state. On the epistemological level, its emphasis on collaborative research and the methodologies of problem posing, discourse, and social learning confront the most pressing and sophisticated epistemological issues facing the social sciences. Underlying these methodologies are critically important questions: How do we analytically integrate empirical and normative knowledge? How do we combine the professional's scientific knowledge with the citizen-client's ordinary knowledge?

Participatory research, to be sure, is not without its problems. Most important is the issue of effective citizen-client participation. How much do we actually know about the ability of clients to collaborate intelligently in technical decision making? Such questions surely require a good deal more exploration and experimentation. The examples offered by popular epidemiology in Woburn and participatory resource mapping in Kerala represent, in this respect, important learning experiences on which we need to build. Not only do such experiences take the idea of participatory inquiry out of the realm of the utopian, but they provide us with a beginning. Both the practice and the theory of participatory research help to clarify the nature of the political and methodological tasks ahead. They help us to recognize, for instance, that many of the problems confronting the reconstruction of professional practices are as social and political as they are epistemological. While there is much to be done in methodological terms, it is at least clear where the search for methodological relevance begins.

Far less clear is the question of where the politics will come from. Much will depend on a commitment to participation on the part of both professionals and the society as a whole. This question emerges as a central issue for the agenda of those concerned with the problems of democracy and expertise.

10. Ordinary Local Knowledge

From Potato Farming to Environmental

Protection

The tradition of knowledge passed from parent to child,
from master to apprentice is the very root of science.
—J. D. Bernal

In chapter 6, I showed that the environmental justice movement, as part of its critique of scientific expertise, emphasizes an alternative ecological perspective grounded in the ordinary knowledge of local citizens. Chapter 8 examined two cases involving local knowledge in environmental politics. For many engaged in these struggles, such alternative knowledge provides citizens' movements with "epistemological tools for the reconstruction of neopositivist science and for an alternative approach to the management of . . . ecological independence" (Breyman 1993, 137).

But what exactly is local knowledge? Is it anything more than a slogan for those who dislike the decisions of the technocrats at the more distant centers of power? Approached rigorously, the answer is not easy or straightforward. As an epistemological concept, local knowledge poses a number of definitional issues.[1] In this chapter, the discussion is limited to one aspect of the concept. Here local knowledge is examined as a specific category of what Lindblom and Cohen

(1979, 12) have defined as "ordinary knowledge." By "ordinary knowledge" they refer to "knowledge that does not owe its origin, testing, degree of verification, truth, status, or currency to distinctive . . . professional techniques, but rather to common sense, casual empiricism, or thoughtful speculation and analysis."[2]

As a subcategory or specific type of ordinary knowledge, local knowledge is knowledge about a local context. Although the production of all knowledge is connected to some degree to a local context, ordinary knowledge more generally can pertain to knowledge of things beyond the local setting; for example, world politics, national unemployment rates, or the operation of nuclear power. Following Lindblom and Cohen, all of these types of ordinary knowledge, whether local, distant, or general, have the same basic epistemological characteristics. In this chapter, ordinary local knowledge refers to knowledge pertaining to a local context or setting, including empirical knowledge of specific characteristics, circumstances, events, and relationships, as well as the normative understandings of their meaning.

Of special importance for Lindblom and Cohen is the role of ordinary knowledge in policy inquiry. Like the hard sciences after which they have tried to pattern themselves, the policy-oriented social sciences have sought to replace traditional or local knowledge with scientifically verified findings. Whereas policy analysis has generally presented itself as an intellectual safeguard against what is taken to be the unsubstantiated opinion of the general populace, the field has failed to recognize its own dependence on such ordinary or everyday knowledge (Schmidt 1993). In a book whose message has largely gone unobserved, Lindblom and Cohen argue that the failure to recognize the methodological implications of policy analysis's dependence on such knowledge is a major reason for the field's inability to supply "usable knowledge." Not only have professional policy analysts overestimated the amount of information and uniqueness of the analyses they offer for social problem solving, but such scholars have also greatly underestimated "the society's use — and necessary use — of an existing stock" including "a flow of new ordinary knowledge from sources other than [professional inquiry]" (Lindblom and Cohen 1979, 12). This chapter first examines in more detail the concept of local knowledge and then turns to its implications for environmental protection.

Recovering Local Knowledge

Commonly described in the past as "traditional" or "indigenous" knowledge, "local knowledge" has tended in recent years to become the accepted phrase. The concept of local knowledge applies to a wide range of human endeavors, from peasant farmers' familiarity with soils, to African hunters' ability to track an animal through a forest, to the botanical knowledge of indigenous peoples, to the rules and strategies of schoolyard basketball. More specifically, local knowledge is the informal, "popular, or folk knowledge that can be contrasted to formal or specialized knowledge that defines scientific, professional, and intellectual elites in both Western and non-Western societies" (Brush 1996, 4). Conceptualized in these terms, "indigenous or local knowledge is the systematic information that remains in the informal sector, usually unwritten and preserved in oral traditions rather than texts." Formal scientific knowledge, in contrast, is organized and carried forward in written texts. Whereas science seeks to theoretically separate its knowledge from the culture in which it is produced, local knowledge remains inherently associated with, and interpreted within, the specific culture in which it is produced.

Thanks to the modern commitment to — if not obsession with — the wonders of science and technology, local knowledge has long been ignored. Indeed, formal scientific knowledge has largely been defined as a superior form of knowledge designed to transcend the limits of indigenous or local knowledge (Schmidt 1993). The very legitimacy of scientific knowledge formally depends on its epistemological differentiation from the everyday knowledge of ordinary people (Lyotard 1986). In many fields, the explicit goal has been to replace indigenous knowledges with more "advanced" scientific and technological knowledges.

The origins of this neglect of traditional knowledge can in part be traced back to the views of early explorers, missionaries, and colonial scientists, particularly their images of progress and the superiority of "civilized" countries. Nineteenth-century principles of sociobiological evolution and "scientific" reason strengthened the belief in the need to convert and improve the "uncivilized savages" of the underdeveloped world and to abolish their primitive, "childlike ways" (Jiggins 1989).

Following in these footsteps, conventional Western scientists have held the beliefs and practices of the underdeveloped world to be based on myth and ignorance and describe them with terms such as "backward," "ineffective," "conservative," "inefficient" (Thrupp 1989). Founded on ignorance, it is the knowledge of "primitive," "stupid" natives and should in the development process be replaced with efficient new expert technologies of the advanced world.

Curiously, in this respect, few observers have noted the degree to which modern science and technology have themselves been built up from the foundations of traditional knowledges. Long before the Age of Science, as Patel (1996, 305) puts it, "people in all parts of the world had been searching for new ways of doing things." Throughout the history of the species, in fact, humans "have relentlessly engaged themselves in the search for more effective tools and instruments to assist their struggle for survival." In retrospect, the resulting step-by-step buildup of usable technical inventions can only be judged as astonishing.[3] Merely consider a brief list of major technical inventions that preceded the industrial revolution: fire, the wheel (followed by carriages and paved roads), the calendar, weaving, pottery, agriculture (including the domestication of animals, irrigation, and the selection and conservation of seeds), arithmetic, geometry, astronomy, sailing and navigation (including the compass), metals and the smelting of ores, gunpowder, rubber, gears, scripts, paper and printing processes, and architecture and city planning, to name some of the most important in the development of modern civilization (T. Gladwin 1970, 1979). This, moreover, is to say nothing of the development of religions and philosophies, states and administrative systems (Moore 1985; Ascher and Ascher 1981).

Seldom do we appreciate how recently the phenomenon of modern science has appeared on the world stage. Well into the eighteenth and nineteenth centuries, scientific and technological innovations and inventions proceeded largely through traditional prescientific methods, including sophisticated forms of alchemy and Renaissance magic. Moreover, these same efforts led to the beginnings of the scientific method, including the search for general principles pertaining to the more immediate problems of everyday life. Scarcely do we recognize either the degree to which the spirit of modern science actually emerges from these earlier efforts, largely trial and error in nature, or the degree to which modern science is merely a formalization of many of the cumulative practices that emerged in "prescientific" times. Indeed, the transition between the prescientific and the scientific is not nearly as sharp as it is often portrayed.

One of the most interesting illustrations of this shading of prescience into science is found in a seldom-told history of Sir Isaac Newton, the single most important contributor to the scientific revolution of the seventeenth century. Little known is Newton's avid interest in the esoteric traditions of occult mathematics and hermetic alchemy, generally considered to be a form of magic (Parsons 1997). In his pursuit of the natural laws of the universe, Newton employed "mystical clues" he believed traceable back to cryptic revelations in Babylon. The British economist John Maynard Keynes, long interested in Newton's scientific manuscripts, made the point this way: "Newton was not the first of the age of reason. He was the last of the magicians."

Some argue that scientific knowledge systems — like all other knowledge systems — are not only made possible by the earlier efforts of others; they are themselves the result of locally based innovations and discoveries. Bernal (1969), for example, sees traditional knowledge as the foundation of science, passed from the master to the apprentice. Stored in the collective memory, often as "tacit knowledge," and passed along in the process of work, scientific or otherwise, such knowledge is part of a long cultural continuum of habituated practices.[4]

Drawing on Bernal's insight into the relation of traditional to modern forms of knowledge, Kurien (1988, 476) illustrates how the traditional local knowledge of the artisan fishermen in southern India should be understood as "practical knowledge which got conditioned by cultural practices."[5] Compared with the modern-day marine biologist's general theoretical knowledge of fishing conditions and practices, he shows the Indian fishermen to possess a vast accumulation of nuanced knowledge of aquatic milieus and behavior patterns of marine life. Passed along from one generation of fishermen to the next, this culturally embedded knowledge pertains to the *diversity* of fish species, oceanographic conditions, and coastlines. He describes the fisherman's knowledge this way:

A fishing operation is not determined *a priori* by a process of inductive reasoning. Any particular fishing operation in progress is a simultaneous integration of a large number of discrete thought processes of past experiences with the immediate observations aided by all the human senses: the feel of the sea-bottom acquired by touching the plumb line; the smell of the sea; the sight of the birds, landmarks, stars, the colour of the sea and the ripples on it; the sound of the shoal movement — to mention a few. The coming together of these aspects initiates the tool using response — dropping of hooks, casting of nets or laying of traps. The result: fish are soon caught.[6] (Kurien 1988, 476)

Official measures by modern fishery bureaucracies to replace this traditional knowledge with the centrally collected data of marine science has often led to policies inapplicable to the circumstances of particular fishing communities.[7] Such policy failures have not only created breakdowns in communication between central ministries and the fishing community but also contributed to commercial crises in overfishing that have disrupted the ecological support systems essential to a steady renewal of fish populations.[8]

Even though such "knowledge systems may differ in their epistemologies, methodologies, logics, cognitive structures, or socioeconomic contexts," write Watson-Verran and Turnbull (1994, 116), "a characteristic they all share is localness." Rather than the linear product of a particular concept of rationality, what we call science is in fact the manifestation of an assemblage of local innovations, technical devices, theoretical languages, practical skills, and social strategies. As we saw in chapter 4, modern scientific findings can be understood as assemblages forged from tensions between local practices and the attempt to translate them into the categories of a global or universal language (Latour 1987; Bourdieu 1988).

Long an interest of the anthropologist, research on local knowledge is closely associated with the study of the knowledge systems of non-Western indigenous peoples and ethnic and minority cultures. Much of our understanding of local knowledge has resulted from the efforts of anthropologists to validate knowledge systems of cultures and languages that have been deprecated and subordinated by dominant national cultures, often threatening their very existence with extinction. Anthropologists, along with linguists, have labored to recover and record the complexity, extent, and usefulness of indigenous local knowledge (Berlin 1992). In some cases, formal systems have been discovered to have their origins in traditional local knowledges (Atran 1987), Western pharmacology and agriculture being important examples.[9] Moreover, as the work of Berlin, Raven, and Breedlove (1973) shows, Western Linnaean botanical classifications are scarcely more systematic than traditional Mayan classifications. There are differences, however. One of the most important is the degree to which indigenous local knowledge is more commonly accessible and widely shared than scientific knowledge (Brush 1996, 5)

Although anthropologists have examined local knowledge for decades (Geertz 1983), there has been a more recent discovery — or perhaps rediscovery — of the knowledge and skills of indigenous peoples by a range of scientists and experts in other fields, in particular agriculture, biology, ecology, and agroecology (McCorkle 1989). Practical knowledge about the environment and agroecology has its origins in the work of Sir Albert Howard, often regarded as the originator of organic agriculture. In a strategy regarded as unorthodox — if not revolutionary — by British colonial administrators, Howard derived many of his ideas by consulting with peasant farmers in India, whom he referred to as "professors" (Howard 1924).

More recently, as seen in the preceding chapter, grassroots activists in Third World countries have recognized ordinary local knowledge based on careful observation and common sense to be a valuable and untapped resource that speaks directly to community problem solving (C. Gladwin 1989; Schmidt 1993). Such knowledge, often generated through methods of participatory research, has been accumulated by local grassroots networks concerned with the quality of the air, drinking water, and tilling soil, as well as with harvesting forest produce and fishing rivers, lakes, and oceans (Breyman 1993, 131). Similarly, Native Americans in northern societies have accumulated firsthand knowledge of the environment unknown to nonindigenous observers. A Canadian study, for example, shows the ways biologists estimating the effect of mega-projects on the ecology of rivers in the northern regions of the country overlooked the existence of various fish species simply because they never bothered to ask native residents who know the land intimately (Richardson et al. 1993, 87).

Such activities can be found as well in advanced industrial settings.

Irwin (1995) describes how British agricultural workers used their own knowledge of local conditions and practices to fight against regulatory rules permitting the use of a highly controversial herbicide with potentially dangerous effects for human health and the environment.[10] Despite repeated assurances from regulatory authorities that there was no evidence to suggest a causal relationship between the pesticide and cancer, among other health problems, the farmworkers presented government officials with a dossier of findings collected from both a range of established medical sources and the farmers' local experiences with the chemical (such as knowledge of spraying conditions, inadequacy of facilities for cleaning and disposal of chemical containers, workers' lack of knowledge of chemical risks involved, distance of the workers from wash facilities, etc.).[11] In addition, the union organized its own database through a membership survey.[12] Such information subsequently provided the basis for a wider campaign about pesticide safety and the regulatory processes designed to protect the health of workers.

Of special significance, as Irwin shows, the data demonstrated the high degree of variability between what the regulators offered as standard operating procedures for use of the pesticide and the actual conditions surrounding local practices with the chemical, which the farmers knew from their own experiences. Moreover, the farmworkers' discovery and recognition of the knowledge and expertise at their disposal had an empowering effect. Complementing the discussion of management and plant workers in chapter 3, the conflict demonstrates the need for a diversity of knowledges relevant to the regulation of environmental risks.

These experiences make clear that the traditional denigration of ordinary local knowledge has been fallacious and naive (Fischer 1995; Richards 1979). What has often been perceived as "ignorance" about advanced Western technologies is not a matter of stupidity but rather a manifestation of poverty, social inequalities, and inaccessibility to resources. In many cases, moreover, the new methods are not adopted because they are unsuited to the needs and environments of resource-poor peoples. In other cases, they have even proven to be inferior to the existing techniques. As more concretely illustrated hereafter, the use of indigenous knowledge, as a logical adaptation to existing conditions and circumstances, is at times a fully rational approach to problem solving.

Local Knowledge in Agroecology

The most developed examples of such local knowledge come from the field of agroecology. In recent years, growing numbers of development professionals and agroecological specialists have expressed positive or laudatory views of local knowledge and capacities (McCorkle 1989; Brokensha et al. 1980). Increasingly, they have described culture-based knowledge with adjectives such as "effective," "efficient," and "functional." Many of these studies refer to technical skills such as cultivation methods or artisanry for tool making. Other studies, however, reveal that such knowledge extends beyond technical aspects and includes nontechnical insights, wisdom, ideas, perceptions, and innovative capabilities that pertain to ecological, biological, geographical, and physical phenomena. In the case of agroforestry, for example, such innovations have been found in pest control methods, multiple cropping patterns, soil fertilization and tilling, polycultures, small-animal husbandry, seed variety, uses of wild plant species, unique botanical taxonomies, and curative herbs, among others.

These insights and adaptive skills of farmers are derived from many years of experience and are part of cultural traditions that have co-evolved with local environments (C. Gladwin 1989). They have often been communicated and learned through family members over generations. Such knowledge systems can include knowledge of various cultural norms, social roles, and physical conditions such as climate or lunar cycles. In some cases, the knowledge is based on its own conception of knowledge, philosophies, understandings, and principles, which differ from modern scientific tenets (Norgaard 1987). In some cultures, the insights are tied to mystical or religious beliefs, or ideas about spirits or ancestral ghosts — which are incomprehensible from the perspective of Western science. Regardless of origins, however, such knowledge often consists of dynamic insights and techniques that change over time through practical experimentation and adaptation to environmental and socioeconomic shifts.

Another feature of such local knowledge is that it is possessed by both men and women (M. Fernandez 1986). In many cultures, women have particularly rich insights about certain resources, plants, tree species, and livestock and farming techniques (Thrupp 1984). For instance, several studies have found East African women to have remarkable knowledge about qualities of different indigenous tree species and their uses for fuel, medicines, and construction (Juma 1989). In some cultures, special information is known only by specific individuals such as midwives, religious leaders, and healing artists (the use of medicinal plants being a case in point).

To be sure, some agroecological studies of local knowledge, as well as popular reports, have too often presented idealistic or romantic visions of such knowledge and capacities (Thrupp 1989). Some convey images of "noble savages," living harmoniously with nature in peaceful states. This view, however, is generally misleading. It should be stressed that the type, extent, and distribution of knowledge varies greatly in Third World societies, as it has in all societies in the past. Similarly, the capacities of individuals to innovate, use, and transfer such knowledge are diverse and related partly to the socioeconomic and environmental conditions of each particular people. Not all resource-poor people have valuable indigenous knowledge; some people have ineffectively relied on beliefs that are detrimental to the people's own interests.

Having acknowledged these points, however, one of the most important signs of the recognition of this local information comes from the efforts on the part of some conventionally oriented social scientists. Growing numbers of analysts suggest that local knowledge constitutes an important source of innovations and skills that can be used and developed for improving agricultural production and upgrading poor people's livelihoods in rural development processes. Some scientists see the knowledge as a potentially important substitute or complement to formal scientific knowledge and technologies (Norgaard 1984). But these efforts largely marginalize or miss the significance of local knowledge, however well intentioned they might be. When research scientists seek to test local knowledge with formal empirical methodologies, using laboratories for controlled tests, they violate the precepts on which local knowledge is founded. Although such studies might demonstrate the validity of local people's practices and ideas, this form of systematization fails to appreciate the true function and the nuances of such knowledge systems. From the perspective of the small farmer, this scientistic abstracting of the people's knowledge by foreign researchers represents a misunderstanding of the farmer's innovative role in the creation of such knowledge. Moreover, the efforts of researchers to form technical "packages" of this knowledge (based on laboratory analyses of species compatibility, etc.), and then to use, transfer, or sell the packages back to farmers, can only be understood as a form of economic exploitation. By perpetuating the usual patterns of marginalization and displacement of indigenous rural peoples, the practice seldom improves the livelihood of these farmers. Understood in this way, the only proper way to obtain this knowledge is to get inside the farmer's system and explore it on its own *socio*technical terms.

Indigenous knowledge, it should also be noted, has in recent years become the topic of a contentious debate. The question has arisen: Who owns and controls this knowledge? Unlike scientific and technical knowledge, indigenous knowledge is not legally recognized; it cannot, for example, be patented. Although the people who develop it freely pass it along to others, agricultural and pharmaceutical companies have sought where possible to codify and exploit such information for commercial purposes, thus giving rise to a major debate about who owns and controls such knowledge. Insofar as existing intellectual property rights systems ignore these contributions, legalizing only the rights of inventors and innovators of modern technology, these legal provisions work to perpetuate the existing inequalities among countries and peoples, rich and poor (Brush 1996). This has led to proposals to treat "cultural and indigenous knowledge . . . as a form of intellectual property in order to increase the economic return from biological resources maintained by peasants and tribal people" (Brush 1996; see also Posey 1990). Given that indigenous peoples maintain large amounts of valuable biological resources, especially resources highly useful to industry, intellectual property is an innovative method for encouraging the development and sharing of knowledge of plants and animal species. It is also a way of protecting both indigenous peoples and biological resources, both of which are endangered. Such proposals are seen to offer a strategy for linking the consumers of these resources to those who maintain them. Biodiversity experts contend that such an approach could lead to land management practices that would help to protect endangered forests, rare plants, and threatened species and encourage crop diversity and the like. Although the future of this issue is not the concern of this book, its emergence serves to illustrate both the validity and the significance of indigenous local knowledge.

Andean Potato Farmers

Local Knowledge in Epistemological Perspective

For a closer look at local knowledge, we can turn to the investigations of van der Ploeg (1993, 1989), who has conducted an extensive study of the local knowledge of Andean potato farmers. Comparing the approaches of indigenous farmers and agricultural scientists, he shows the ways in which the local knowledges of potato farming, contrary to conventional opinion, are a dynamic system of complex knowledge, involving sophisticated reflection on both material factors and cultural values. Whereas agricultural scientists work with standardized laboratory conditions, the local knowledges of the farmers are inherently interwoven with the immediate practices of their craft. Working from the presumption of an ideal or optimal seed type, coupled with standardized environmental conditions, scientists seek to define, test, and redefine the proper combinations of seed and plot characteristics in experimental stations.[13] In sharp contrast, the potato farmers work from actual conditions. They select seeds according to the variable climatic and environmental conditions of the land, a process that they continually monitor and adapt over an extensive time. Involving ongoing assessments of their production experiences, the deliberations of the farmers bring together the interactions of both manual and mental labor. The result is a well-developed system of agricultural knowledge coupled with an adaptive multidimensional culture designed to deal with unknown and varying situations. Rarely expressed in a clear, univocal form, the concepts of such knowledge are not unequivocal and do not lend themselves to precise measurement and quantification.[14] As they cannot be fitted into a nomological model of the kind used in applied science, they seldom lend themselves to the standard methodologies of management and planning.

Most typically, these "unscientific" characteristics of local knowledge are enumerated to denigrate and dismiss it. But such knowledge should in no way be rejected as contentless. Even though such knowledge is linked to pragmatic work practices, careful examination, as van der Ploeg shows, reveals something akin to a theoretical structure. To be sure, it presupposes none of the general characteristics required to be categorized as scientific in the conventional sense of the term: it possesses no system of impersonal values or a systematic conceptual structure. Indeed, when detached from the people who use it, such knowledge can easily be judged as "unusable," even "inaccurate" and "unreliable." For the farmers, however, it is just this "inaccurate" or "irregular" character that permits them to establish quite accurately the overall conditions of specific fields. Indeed, their knowledge derives its value from this flexibility and openness to uncertainty. Its variability is essential to their conscious effort to build diversity into their practice of agriculture. Or stated the other way around, this flexibility is a manifestation of their age-old struggle to cope with a multiplicity of factors involved in the not so simple effort to eke a living out of the ground. Rather than seeking to standardize the environmental conditions of the land, as do agricultural scientists, the potato farmers do just the opposite: they actively work to increase the variety of conditions. Operating with no fixed conception of optimality, the farmers take multidimensionality and variability as givens. They creatively employ their knowledge of such variations to inform practical assessments of existing conditions and potential improvements.

Moreover, not only do these farmers communicate with one another about these assessments, but the variability and flexibility of their concepts serve as the currency of their collective assessments. As van der Ploeg (1993, 212) explains, the very "inaccuracy" of their concepts renders them "favourable for such an exact interpretation of a plot's conditions and the ensuing dialogue." Such communication and interpretation, he writes, "can only be active processes [in which] concepts must be weighted against each other every time a specific plot is being considered" (212). Thus "the conceptual overlap becomes strategic" (212). It is exactly "the vagueness or 'imprecise' character that allows for this active process of interpretation and change" (212).

Because local farming knowledge does not lend itself to codification, as van der Ploeg (1993, 220) writes, neither "the outcome of such methods [can] be exactly predicted . . . nor can the necessary methods for reaching pre-established levels be prescribed in detail." Although this poses no difficulties for the farmers, it is a fundamental problem for the scientists. Despite the advantages indigenous methods can offer in the context of varied circumstances, the fact that they lack the regularities necessary for standardization makes it difficult, if not impossible, to integrate them into scientific practices. Local knowledge and methods are thus incompatible with the very logic of scientific design. As a result, "farmers as active and knowledgeable actors, capable of improving their own conditions, also fall outside the scope of scientifically managed rural development" (220).

Van der Ploeg concludes that scientific modernization programs not only destroy local knowledge but render the farmers themselves impotent. Basic here is a cycle of dependencies that scientific farm management sets into motion. For the potato farmers and their way of life, the scientists' efforts to optimize their outputs through standardized mass production involve a negation of the farmers' cultural knowledges and practices.

Insofar as scientifically engineered seed types require standardized field conditions that can be repeated in each field, the process means more than the mere use of the new seed type. For the farmers, it involves a sophisticated reorganization of farming routines that has no place for their knowledges of diversity, complexity, and methods of local learning (van der Ploeg 1993). Indeed, one could argue that the very goal is to reorganize the farmers' knowledge out of the process. By averaging together conditions in various, often quite different fields, the scientists create a single quantitative value to be uniformly applied to each plot. Lost in the process is the farmers' ability to enumerate a range of differences across these fields, from soil characteristics and cultivation practices to the climatic effects of wind and frost. Nobody asks them; the differences are simply averaged out.

For the potato farmers, the rejection of their expert knowledges is more than a matter of agricultural optimization; it is a threat to their very social and cultural identities. In no uncertain terms, the standardized procedures of scientific farm management involve a transfer of social agency from the farmers to the scientists. In the process, specific quantities of fertilizer are required to avoid damaging the crop. This involves spreading specific chemicals according to exact timetables experimentally established in terms of the seed type properties. This in turn means following specific irrigation requirements and a host of other contingent practices. Moreover, these prescribed requirements must be repeated systematically in the fields over specific periods of time. Otherwise, for scientific management, it means project failure (van der Ploeg 1993, 219; Wynne 1996, 71). For the farmers, it means replacing their own agricultural savvy with the manager's standardized book of procedures. Experienced as a negation of their identity as skilled craftsmen, the farmers increasingly find themselves alienated from scientific procedures as a whole.

Stated more critically, the scientific system "works" only if it can exclude large parts of the social world from its purview — not least important the fact that noncodified systems of indigenous expertise are themselves capable of producing quite respectable yields. Observing this ignorance of their methods and practices, farmers recognize this neglect on the part of the scientific community to be not only a precondition for the introduction and development of scientific farming but a practice undertaken in the interest of outsiders. Scientific farm management tends at a point to start working against itself. In its effort to replace indigenous knowledge, as van der Ploeg (1993, 22) writes, science "produces just the opposite effect, at least under the circumstances described: myths, vagueness, poly-interpretability and a certain subjectivity in the relations to nature are not superseded through heavy inputs of applied science, but rather reinforced and extended to farmers's relations to science itself." Science becomes a demon perpetrated on them by evil forces from an outside world.

Deriving its particularities from space and time, then, traditional local knowledge is largely utilitarian in nature. When its information, ideas, and concepts are adaptable to the local culture and its technological base, they are integrated into the local social epistemology (Krimsky 1984). But this knowledge, as Chambers (1981) points out, is not merely factual in nature. Such local knowledge also frequently involves a tacit awareness or understanding of the complexity of the agroecological system as a whole. Compared to scientific knowledge, such conceptual systems select different parts of the world for identification and demarcation. In important respects, these contextual demarcations offer a valuable check against scientific tendencies to emphasize generalized knowledge. But this local knowledge offers more than just the possibility of filling gaps in a scientific paradigm. It also involves alternative ways to conceptually organize and understand nature.

As in the case of the potato farmers, most of the literature on local knowledge tells rather tragic stories about the loss of an important resource. Beyond recognizing the validity of an endangered species of knowledge, however, there are positive experiences that illustrate local knowledge's uses in contemporary contexts, particularly in integration with modern practices. In this regard, the next section turns to a South African example that demonstrates how such knowledge has been put to good use in the management of public game parks.

Using Indigenous Knowledge
Wildlife Conservation in South Africa

In contemporary South Africa, where the political regime established by Nelson Mandela confronts the task of transforming a highly segregated social structure into a multicultural society, one of the ways to respect and integrate the many tribal communities has been to acknowledge their indigenous cultures and knowledges. Indeed, whereas the concepts of local knowledge and participatory research have little familiarity in the modern industrial societies of the North, in a country such as South Africa, they are literally among the buzzwords. At the University of Witsvatersrand in Johannesburg, the School of Public and Development Management has sought to integrate these concepts into the curriculum alongside traditional planning techniques and statistical methods. Drawing on critical methodological discussions, the school has attempted to cast its efforts in terms of a postpositivist understanding of the social sciences. Similarly, the South African magazine *New Ground: The Journal of Development and Environment* was established to offer development and environmental activists, as well as other professionals, practical examples of indigenous knowledge and how it might be used alongside standard methods.[15]

One important example of these efforts has been in the field of wildlife conservation, a critical topic in South Africa. Not only is wildlife preservation an inherent part of traditional South African culture, but it is one of the country's most significant resources. In addition to being a critical source of tourist revenues, it is an important asset for species preservation and tourist revenues. For this reason, the problem of animal poaching in the country's reserves has been an especially troublesome issue. Given the lucrative trade in elephant tusks and animal skins, a major effort has been undertaken to protect wild animals against the large number of illegal hunters who track and kill them for personal gain. While the most serious threat to animal preservation is posed by commercial poaching syndicates, the temptation to poach is also strong for the more adventurous of South Africa's large poverty-ridden population, perhaps 80 percent of the country's citizens.

This latter issue has been especially exacerbated by the antipoaching efforts of past governments. In an effort to combat commercial poaching, such regimes have passed antipoaching laws that have eliminated subsistence hunting for indigenous communities as well. In the process, a legitimate use of wildlife resources became illegal, depriving indigenous communities of a resource that they had traditionally managed for centuries. With legal access to wildlife denied to them, survival has often meant that indigenous hunters, desperate to feed their families, have had to "steal" game from the government. Some, moreover, have accepted lucrative offers from commercial poaching syndicates to do their stealing for them. Assuming the heavy risks involved in poaching rhino and elephant tusks — in some cases "shoot on sight" — the hunters typically receive little more than subsistence wages for their efforts. The commercial smugglers, reaping huge profits on the international markets, almost always escape capture and prosecution.

Despite shared concerns, the efforts of the government's conservationists to intervene effectively in the poaching process have been disappointing. It is widely agreed that national park officials know neither how many animals are being killed nor how many people are doing it. Although the park conservation departments employ trained "specialists," few prove to be especially skilled at tracking animals. Most are former military personnel trained to track humans rather than wild animals. Equipped with radios, jeeps, helicopters, and computerized information on animal movements, these wildlife rangers have been unable to get a firm grip on the problem. As the head of one antipoaching unit explains, his former army trackers "can walk right past a rhino carcass in the thick bush and not know it. They don't know to look for jackal and hyena tracks to tell them where there is a carcass nearby. As a result, there are no accurate figures of how many rhinos are being killed" (Liebenburg 1993, 24).

These failures have led some tracking units to adopt an alternative strategy. In place of conventionally trained trackers, they have turned to indigenous hunters with native tracking skills. Replete with references to traditional mythologies, the understanding of these native trackers differs from that of the zoologists. Whereas zoologists and conservationists work with computer-based predictions about the location of animals, native trackers make predictions on the basis of what can best be described as trained intuition. Combining logic with intuitive sight, the traditional tracker does more than just observe an animal's behavior. In following a trail, the tracker puts himself in the place of the animal and asks himself what he would do if he were that animal. Literally projecting his indigenous ways of thinking onto the animal, the tracker works through an intense empathetic concentration to actually "feel" like the animal. Such trackers maintain that they can physically and emotionally sense when an animal is nearby. According to one observer who has accompanied such trackers during their work:

> !Xo trackers say that when a dangerous animal such as a lion is nearby they feel a burning sensation in the middle of their foreheads or when an antelope is near they feel a burning sensation in the armpits. No matter whether these forebodings are real or not trackers have faith in them. And when you are tracking a lion with them, armed only with spears and throwing sticks, you certainly hope they are right! (Liebenburg 1993, 26)

Such behavior can easily be criticized as anthropomorphic, but it might not be as unscientific as it first sounds. To the contrary, there may well be some basic connections between intuitive measures employed by the trackers and the kind of imaginative speculation long recognized as common to scientific endeavors. Most scientists would take umbrage at the suggestion, but to make the point, Liebenburg (1993, 26) cites the Nobel Prize–winning French biochemist Jacques Monod, who argues that a theoretical physicist should at times identify himself or herself with a particular nuclear particle and ask what he or she would do if he or she were that particle.[16] How would I behave? Where would I move? Even though the comparison might be overdrawn, it resonates with postpositivist epistemology.

Further support for such an argument can be found among zoologists and ecologists. Increasingly, such scientists have themselves turned to indigenous experts for assistance in studying animal behavior. In a study of the possible effects of rhino de-horning, for example, scientists discovered that traditional tracking is an invaluable supplement to radio tracking. The physical tracks and other signs of animal behavior reveal a great deal of valuable information about a species's everyday activities that otherwise goes undetected. Finding it advantageous to combine such informal information with their scientific models, some animal scientists have begun to think more systematically about the relationship between the two types of knowledge. Although the zoological models of animal behavior are in many ways more systematic, zoologists have come to recognize other ways in which the indigenous trackers' knowledge is broader. For hundreds of years, such trackers have been aware of both general and specific characteristics of animal behavior that Western scientists have only recently discovered. In recognition of this, some zoologists, ecologists, and conservationists now speak of the need to appreciate the value of instinctive knowledge, something long understood by anthropologists who have studied the same hunter-gatherer societies (Liebenburg 1995).

Although the question of the epistemological status of such intuitive knowledge is not so easily settled, the point here is only to indicate that there is more to the issue than is normally recognized or acknowledged. What is clear is that for the conservation of wildlife, the use of traditional trackers has proven advantageous. Moreover, it can be seen as part of a larger strategy to bring together wildlife conservation and the interests of the community. If community involvement is an essential element of a strategy to control poaching, as numerous experiences suggest, the engagement of local trackers is an excellent way to begin building in the interests of the community. Creating local conservation jobs forges a vital link between employment opportunities and community involvement. Instead of using conservation monies to buy helicopters (mainly from First World countries), the money spent on trackers goes directly into the local community, providing the community with an added incentive to conserve wildlife. By maximizing available human resources, the local economy is stimulated at the same time that wildlife protection is made more sustainable.

Along these lines, some have suggested an even broader community-based approach to antipoaching efforts. The first step would be to legalize local subsistence poaching (which goes on anyway, usually with little detection). Allowing controlled subsistence hunting in conservation preserves brings the hunters into the process of wildlife management. Subsistence hunters could then employ their indigenous knowledge and skills not only to track animals for a livelihood but also to provide information helpful in tracking down commercial poachers. Instead of working for the commercial smugglers, the hunters would become involved in the effort to track them down. Some argue that this is the key to bringing conservation efforts to a new level.

While there have thus been some positive developments in recognizing and employing trackers, there is still much that needs to be done to ensure that this form of local knowledge is preserved for the future. At present, the small number of indigenous trackers employed in conservation jobs constitutes the last of an age-old tradition. Few resources are provided for developing the socioeconomic conditions that work to support the survival of local knowledge and expertise. For instance, civil service rules make it difficult to employ indigenous trackers. While the local hunters possess superior tracking skills compared to conventional "specialists," they can seldom meet civil service employment requirements. To be employed by the government as a tracker, one must have earned a school certificate, be able to use a radio, read maps, drive a vehicle, and ride a horse. Many, however, cannot read and write, let alone possess a school certificate.

Ordinary Local Knowledge

Implications for Policy Inquiry

What does this obvious presence and importance of local knowledge have to do with environmental policy analysis? If we follow Lindblom and Cohen, there is no alternative to relying significantly on ordinary local knowledge. The basic information necessary for political action takes the form of ordinary knowledge, much of it local in nature. Moreover, policy science has offered little or nothing to replace it. In fact, ordinary knowledge is mainly what policy analysts themselves have to offer. It is mainly the possession of large amounts of high-quality ordinary knowledge in a particular field that qualifies them as experts. Indeed, were the title "expert" conferred for the possession of scientifically verified knowledge alone, given the small amount of such available knowledge, the application of the term would be at best ambiguous.

To be sure, the social sciences have tested and refined important beliefs grounded in ordinary knowledge. But the actual amount of such knowledge that has been tested and refined is little more than a drop in the bucket. Moreover, on those occasions when policy analysts do possess such knowledge, the advantages of such knowledge can seldom be taken for granted. As Lindblom and Cohen (1979, 13) put it, "just what the advantages are needs to be identified" (see below). Policy-analytic knowledge provides neither an adequate basis for social problem solving nor sufficient evidence to support the belief that policy research holds out the possibility of replacing ordinary knowledge as the currency of decision making. Indeed, J. C. Scott (1998) provides a useful analysis of some of the major failures — even tragedies — resulting from a naive application of such knowledge.

Professional Knowledge and Ordinary Knowledge

Lindblom and Cohen offer the following points to clarify the relationship of professional social inquiry (PSI) to ordinary knowledge.

1. For a pPSI [practitioner of professional social inquiry], we suggest that most of the knowledge which appears in his work is ordinary knowledge that is widely dispersed among relatively informed members of society and not the product of PSI . . . verification. That businessmen will not invest if earnings are not anticipated, that in many circumstances prices and wages spiral upward together, that disciplinary problems in schools distract teachers from educational work . . . and that television absorbs much of the time of children are widely known (or believed) propositions. Of many important propositions like these, pPSI knows no more than

many other informed members of our society. Nor can they usually offer more verified versions of such propositions than others can, although they can provide some degree of verification for a few of the many propositions they present.

2. Moreover, much of the "new" knowledge produced by PPSI is ordinary knowledge. That is to say, it is produced by the same common techniques of speculation and casual verification that are practiced throughout the society by many different kinds of people, and is not by any significant margin more firmly verified. Despite the professional development of specialized investigative techniques, especially quantitative, most practitioners of professional social inquiry, including the most distinguished among them, inevitably rely heavily on the same ordinary techniques of speculation, definition, conceptualization, hypothesis formulation, and verification as are practiced by persons who are not social scientists or professional investigators of any kind. . . .

3. When practitioners of PSI push on to new knowledge significant for social problem solving, they can produce, we suggest, no more than a small number of propositions — tiny compared with the stock of propositions commonly employed in social problem solving. That is to say, their offered increment of new knowledge is just a veneer or, again, an addition to a mountainous body of knowledge.

4. We further suggest that part of the new PSI knowledge generated by investigatory techniques distinctive to PSI is an even smaller flow, perhaps characterized as a trickle.

5. The distinctive techniques of PSI, we suggest, are more often used to test existing propositions growing out of and circulating as ordinary knowledge more . . . than it creates new previously unformulated knowledge.

6. When it refines it does so selectively — indeed highly selectively. The number of propositions drawn from ordinary knowledge that are subjected to distinctive PSI forms of testing represents an extremely small fraction of the knowledge used by PPSI in their work and of knowledge for social problem solving. And as for propositions thus tested, which as a consequence can be said to have been given a high degree of verification, their number is extraordinarily restricted. One is hard-pressed to think of examples.

> Consequently, PSI is not broadly a distinctive source of information and analysis; it is only occasionally so. In saying this, we do not intend to slight the effects that PSI sometimes appears to exert on social problem solving. We quickly acknowledge that the veneer of and the occasional verification of knowledge may often be critical, crucial or pivotal. We do, however, want to suggest the relation of PSI to a mountain of ordinary information that it cannot replace but only shape here and there. (Lindblom and Cohen 1979, 14–16)

Thus the idea that the social and policy sciences are building a "ground-up replacement of ordinary knowledge by scientific knowledge" can scarcely be taken seriously. The basic disciplinary ideology that professional practitioners offer a superior form of knowledge available only to those trained in the discipline's techniques requires serious qualification. Although professionals do possess more specialized knowledge than lay citizens, it is scarcely as exclusive as we are generally led to believe (Schmidt 1993). Indeed, "many journalists, civil servants, businessmen, interest group leaders, public opinion leaders, and elected officials" are fully knowledgeable on such subjects as well (Lindblom and Cohen 1979, 13). In some cases, even interested citizens can have such knowledge.[17] This is especially likely when an issue turns on local contextual factors, as seen in this chapter. In fact, it is just this distribution of knowledge and the many possibilities for disagreement that it creates that have given rise to the "politics of expertise" in the realm of public policy (Fischer 1990).

In short, contrary to their claims, professional analysts have no *unquestionable* knowledge advantage. Indeed, the limited supply of what are agreed to be scientifically verified findings, as discussed in chapter 4, raises important epistemological questions. In the face of these practical limitations, Lindblom and Cohen (1979, 18; see also Lindblom 1990) suggest that policy analysts "study and develop possible cooperative relationships with other sources of the ordinary information on which they themselves heavily depend." Local knowledge, as we have seen here, has to be one of them.

As a step toward such investigation, Krimsky (1984) offers a useful outline of potential nonexpert contributions to technical knowledge.

In the area of problem identification, he points to lay knowledge of an existing problem, diagnosis of environmental or social pathologies, and suggestion of causal connections and explanatory hypotheses. As examples, he offers cases similar to that of popular epidemiology in Woburn, whereby local citizens identified both the presence of disease and the toxic contaminants responsible for it. With regard to evaluative understanding, he points to the knowledge of cultural elements necessary for social cohesion, the value and meaning of social and physical forms, and community needs. Krimsky offers experiences similar to that of participatory resource mapping in Kerala, where community members explain the evolution and social meaning of various existing local structures and activities to the professional planners. For particularized or contextual knowledge, he indicates knowledge of particular events, direct experience and personal knowledge of the local circumstances, and appreciation of environmental complexity. Krimsky illustrates the contribution of this kind of local knowledge through the experiences of farmers, such as the indigenous potato farmers in the Andes, with their special awareness of the agroecological circumstances necessary to successfully harvest their crops.

In the context of specific policy practices, Krimsky (1984, 252) illustrates the need for nonexpert knowledge in the field of urban planning. Concerned with the design of the physical environment, architecture and urban planning have been dominated by overly rational, deductive modes of reason. In their design practices, they make important decisions about the social and cultural significance of the functions and form of physical structures. Toward this end, they bring to bear "the canons of their profession, the architectural fashions of the day, the desires of their clients, and their personal values." In the process, such planners have largely neglected the ideas and interests of the community's ordinary citizens. As Krimsky points out, however, "those who have an intimate connection with the urban environment, who have walked and played in the streets and shopped in its stores, are uniquely qualified to discuss the meaning and value of its physical forms." An urban community, like a sensitive ecological system, evolves over a substantial period of time. But as many of the failures of urban planning have made abundantly clear, the social and physical character of the local landscape can be altered irreparably by professionals who fail to learn from those with an intimate everyday relationship to the urban environment. As planners such as Jacobs (1961) and Friedmann (1973) have shown, the citizens' local knowledge of community life can and should make the professional planner aware of the basic social meanings and values underlying the existing structures and forms.[18] To counter the rational model of planning and its failures, Friedmann has called for a relation of "mutual learning" between professionals and citizens. Only through professional practices capable of combining particularized and generalized knowledge of the urban environment can the planning process make a meaningful, lasting contribution to development of viable city life.[19]

Conclusion

This chapter has more explicitly examined the nature of local knowledge, the primary objective of participatory research. Both the case of the Andean potato farmers and the case of the South African trackers reveal the importance of the local contextual knowledge of local inhabitants for effective social action. As countless implementation studies show, the application of any principle or policy to a specific normative context requires an assessment of the empirical circumstances of the situation (Fischer 1995). Beyond such empirical considerations, local knowledge plays an important role in problem identification, definition, and legitimation, not to mention any solutions that may be put forward. Insofar as many social problems have their origins in a local context—environmental problems being a prime example—knowledge of the local citizens' understandings of the problem is essential to effectively identifying and defining the problem. Moreover, a good deal of experience shows that citizen involvement in both defining a problem and searching for its solutions is an important factor in building the legitimacy required to implement policy effectively. Without the understanding and consensus of the local actors, as many failures have shown, the chances of successful policy intervention are less than encouraging.

We have also seen that local knowledge can have its own epistemological standing in relation to empirical science. Not only does local knowledge have its own epistemological status as a form of informal knowledge, but it has an important role to play in both empirical and normative analysis. For empirical analysis, it speaks to the particular circumstances of a situational context that is ignored by the positivist search for universal principles. Indeed, the goal of positivism has been to seek universal knowledge independent of social context; that is, to discover empirical principles that apply to all contexts regardless of specific circumstances. As we saw in chapter 4, the idea of such knowledge rests on a false understanding of social reality. Before continuing with these epistemological questions, however, I explore the larger implications of local knowledge and citizen participation, particularly as they pertain to societal-level policy decisions.

PART IV

Discursive Institutions and Policy Epistemics

Majority rule, just as majority rule, is . . . never *merely* majority rule.
[It means] antecedent debates, the modification of views to meet
the opinions of minorities. . . . The essential need, in other words,
is the improvement of the methods and conditions of debate,
discussion, and persuasion. That is *the* problem of the public.
—John Dewey

The final part of the book seeks to set the issues of participatory
inquiry in the broader political and epistemological contexts of public
policy and policy analysis. Chapter 11 turns to the question of participation as it pertains to institutional practices and professional conduct. It examines in particular the ways in which participatory inquiry
can be extended beyond local issues to societal-level questions. Toward this end, the chapter begins with a discussion of the concept of
"civic discovery," which serves to anchor the concerns here to contemporary debates in the emerging theory of public management advanced by writers such as Reich and Moore, including its extension to
environmental policy. Although this theory emphasizes the need for
facilitation of public discourse, the proposed practice of civic discovery is seen to fall short of a truly participatory model.

Civic discovery, as will be argued, would make available better information and arguments for the processes of policy formulation, but
it remains unclear how it would bring citizens closer to these processes. In the interests of democratic participation, the discursive role
of the public manager needs to be structured more specifically as that

of facilitator of *citizens'* deliberation. Toward this end, the chapter suggests the Danish "consensus conference" as an alternative model for rethinking the possibilities of building into public administration a more participatory approach to citizen deliberation on complex societal questions. The consensus conference provides a model for invigorating democratic practices. It also serves to build trust and understanding among citizens, policy experts, and decision makers. As such, the consensus conference would appear to be a suitable institutional model for the democratization of civic discovery.

Chapter 12 closes the book with a discussion of the prospects for a discursive model of policy inquiry. The development of a deliberative model, in particular its emphasis of the role of the expert as facilitator of policy discourse, needs to be accompanied by a new subspecialization of "policy epistemics." Concerned with the nature of communicative relationships among different groups of inquirers, especially citizens, experts, and policy decision makers, policy epistemics focuses on the ways policy knowledge is constructed and framed, disseminated, and interpreted. Of special importance, it involves innovating methods needed for coordinating multiple discourses in and across institutions. The chapter concludes with a discussion of the implications of policy epistemics for the expert as facilitator of citizen deliberation.

11. Discursive Institutions for Environmental Policy Making: *Participatory Inquiry as Civic Discovery*

The tradition of democratic deliberation, with its emphasis upon what is good for *society* . . . has been subordinated in part . . . because of our culture's understandable fear of demagoguery and intolerance. . . . But there may be greater danger in failing to appreciate the power of public ideas and the importance of deliberation about them. In an era like the present one — when overall public purposes are less clear than during wars or depressions . . . when many issues are so technically complex that values are easily hidden within expert judgments . . . our strongest bulwark against demagoguery is the habit of critical discussion about and self-conscious awareness of the public ideas that envelop us.
— Robert B. Reich

It is common for many to ask how realistic participatory inquiry might be in the existing political world. This is especially the case when one speaks of participatory research in the kind of Third World liberation language employed by writers such as Freire. Even if it might be good for small groups of indigenous peoples dealing with issues such as literacy and farming, how can it be relevant to the larger, complex policy questions facing advanced industrial nations? It is to this question that this chapter is devoted.

The question would have been more difficult to answer ten or fifteen years ago than it is today. Since then, emphasis on deliberation has become an important theoretical topic in disciplines such as political science, public administration, and planning. Furthermore, these theoretical concerns have begun to make their way into practical policy discussions. In the context of environmental affairs, for example, the National Research Council (1996) published a report calling for participation and deliberation in risk assessment. As we see in the next chapter, this new deliberative orientation is an important step forward for an organization that initially prescribed a technocratic approach to risk assessment and management. Today, deliberative participation is not only seen as a normative requirement for a democratic society but serves increasingly as a counter to the uncertainties of science. As such, it is recognized as the way to build trust in an environment of uncertainty.

What is more, discourse has emerged as a liberal reform strategy for the public sector more generally. One of the more prominent theories in recent years suggests that the role of public management should be recast in terms of the organization and promotion of public discourse. Extending this approach, the first half of this chapter argues that if taken seriously, such a position would indeed involve the introduction of structures and practices quite similar to the kinds of participatory practices discussed in the previous chapters. The second half of the discussion suggests a model for such practices. Already in the preceding three chapters we have seen how the expert can function as a facilitator. The illustrations there, however, were limited to inquiry about concerns inherent to particular local contexts. In this chapter, we examine the possibility of such modes of inquiry dealing with broader technical issues of national and international concern. Although not yet widely recognized, they have indeed begun to take shape, especially in northern Europe.

Public Institutions in Crisis

There is no shortage of discussion in Western democracies of the need to reform modern institutions and their decision-making processes. This is especially the case for governmental administrative institutions, which have long been a favorite scapegoat for all that is seen to be wrong with modern government. There have been a variety of reform strategies in recent years. The most prominent strategy—especially in the United States and Britain—has been a call for the elimination or drastic reduction of government agencies and their replacement with free-market approaches. This strategy, however, is as much rhetoric as it is reality. Even a casual look at the experiences of those advocating such an approach shows it to defy the realities of modern government, whether run by liberals or conservatives. Conservatives, despite their professed agenda, have mainly only made big government bigger.

Many liberal politicians and intellectuals, on the other hand, acknowledge the "crisis of the state"—to which they have significantly contributed—but recognize that administrative institutions must remain a central part of the political process. Rather than getting rid of government administration, the question for them becomes how to reform it. In this view, one of the basic issues concerns the professional orientation of public managers. The crisis—whether concerning aid to the poor or environmental protection—is seen to be caused in significant ways by the expansion of professional authority in solving public problems. Not only has professional dominance brought about client dependencies, but many of our most pressing problems have proven intractable to instrumental forms of technical reason. Moreover, professional practices have failed to attend to the requirements of democratic governance, in particular the need for self-governing communities with responsible citizens managing their own affairs. As Sirianni and Friedland (forthcoming 2001, 101) explain, the policies of public and nonprofit agencies are professionally driven by "a deficit model that focuses on specific problems and needs rather than aiming to build upon the individual associational and institutional assets and networks already existing within these communities." Along the way, public institutions that "formerly played an important role in educating for broad citizenship responsibilities have transformed themselves into narrow service organizations" (101). Coupled with increasingly balkanized communities, particularistic claims, and consumeristic cultures, the result has been an erosion of a sense of common identity rooted in shared public work among American and European citizens.

In the face of these problems, the progressive alternative has been to rethink and reconstruct the administration of the public's business. Rather than disbanding public agencies, the strategy is to decentralize and democratize them. In this view, there is a need for public policy for democracy itself—that is, policies and programs designed to enlighten, empower, and engage citizens in the process of self-government (Ingram and Smith 1993). Such policies would work to promote the search for new ways to strengthen civil society and community capacity. Basic to such policies is an emphasis on deliberation around common interests and concerns.

Public Administration as Deliberative Policy Making

While a good part of the call for renewal has come from an active civic reform movement, one of the most interesting approaches to this deliberative strategy has taken shape in public administration theory. Of particular importance has been work at Harvard University's Kennedy School. Writers such as Reich and Moore have sought to spell out a new theory of public administration, or what they prefer to call "public management," grounded in the practices of public deliberation. Although their work primarily addresses the question of administrative power in the modern state, it speaks at the same time to the task of reorienting the role of the professional expert vis-à-vis the citizen.

Fundamentally, Reich and Moore take up an old issue that has become especially chronic in the era of big government; namely, what to do about the fact that public agencies have become primary policy-making arenas. That is, as Congress and the presidency have delegated more and more discretionary policy-making authority to public agencies, the venerable theory of "overhead democracy" has lost credence. No longer does the conventional understanding of the public agency's role as merely efficiently executing the decisions of lawfully elected officials explain public administrative behavior. As a result, the administrative state has dramatically expanded, but without a theory to explain and legitimate its apparent political role in the policy-making process.

Much of the debate about administrative discretion has thus been tied up in the long-standing "politics-administration dichotomy," around which much of public administration theory has moved. The counterpart of the separation of facts and values in positivist epistemology, the dichotomy spells out a strict division of political and administrative decisions. In the ideal model, often referred to as the top-down model of politics, political decisions are made by the legislature, which then passes them along to the public agencies for efficient administration. The administrative emphasis on efficiency, moreover, has in large part been the basis for a rigorous positivistic approach to public management.

But the model has never really corresponded to reality. Political behavior has always been a significant part of administrative behavior and has become ever more so as government has become bigger and more complex. The question has been what to do about this. One position, traditionally identified with Herbert Finer (1941) but later touted by such luminaries as Theodore Lowi (1979), has held that this particular form of politics—namely, the granting of administrative discretion—should be narrowed and reduced to the smallest degree possible. Proponents of the dichotomy have argued that a line can and should be maintained between the politics of policy making and the execution of policy, that is, the administrative function. The proper role of the civil servant is to execute policy decisions, not formulate them.

The opposing view has always seen this dichotomy as a misleading distinction that has done more to mystify the nature of public administration than to inform it. In this view, the formation and execution of public policy constitute continuous processes that cannot be separated (Friedrich 1940). Today most scholars take the limits of the dichotomy for granted. Involved in much more than the execution of legislative decisions, public managers are seen to be constantly and inescapably involved in political actions (Lynn 1987).

But accepting this reality has done little to solve the traditional problem. If public managers are involved in political work, how is their role to be legitimated? No principle is more basic to democratic government than the idea that political decisions are made by politically accountable officials, the electoral process being the primary means of insuring that accountability. How then are unelected public servants to justify their part in the process? The standard answer has been that they are accountable to the elected officials, that is, those who can replace them. But as government has gotten bigger and more complex, this strikes most scholars as implausible. To ask the president of the United States to be accountable for the actions of thousands of administrators, most of whom he has never met, is largely an unworkable solution.

For Robert Reich (1990, 1988) and Mark Moore (1995, 1983), the prescription for this problem of administrative discretion is more, rather than less, political engagement, at least understood in terms of public education and deliberation. In Moore's words, it is "inevitable and desirable that public managers should assume responsibility for defining the purposes they seek to achieve, and therefore to participate in the political dialogue about their purposes and methods" (M. Moore 1983, 2–3). The most interesting and important version of this theme has been put forward by Reich, former Harvard Kennedy School professor and secretary of labor in the Clinton administration. The central question Reich (1988, 138) seeks to address is how to find a method that can "inspire confidence among citizens that the decisions of public managers are genuinely in the public interest." His advice is to reject public manipulation and replace it with a deliberative relationship with the citizenry (Reich 1990). Reich argues that public managers must assume a more active role in fostering public deliberation. In his view, the goal of deliberation is to build legitimacy for policy decisions ultimately made by public officials. Instead of only making and implementing policy and decisions, as he explains, the task is to organize and manage a process of public education and deliberation, or what he calls "civic discovery" (1988, 144).

Civic discovery as deliberation, according to Reich, refers to a "process of social learning about public problems and possibilities" (Reich 1990, 8). The goal of deliberation is the "creation of a setting in which people can learn from one another." As an ongoing and iterative process requiring two-way communications, such deliberation focuses on "how problems are defined and understood, what the range of possible solutions might be, and who should have the responsibility for solving them" (Reich 1990, 7).

To be sure, public executives in this model bring certain ideals and values to the process, even specific ideas about what they think should be done. But most important, they look to the public and its intermediaries (e.g., citizen groups, the press, and other government officials) as sources of guidance in setting direction. In addition to being straightforward about their own values and perspectives, they form their agenda only after listening carefully to the deliberations of others. In the process, public managers do more than simply seek to discover what people want for themselves and then attempt to find the most effective means for satisfying these wants. Their task is also "to provide the public with alternative visions of what is desirable and possible, to stimulate discussion about them, to provoke reexamination of premises and values, and thus to broaden the range of potential responses and deepen society's understanding of itself" (Reich 1990, 9). Rather than resting on "thoughtless adherence to outmoded formulations of problems [policy making] . . . should entail the creation of contexts in which the public can critically evaluate and revise what it believes" (Reich 1990, 8).

Reich concedes that deliberation is time-consuming and uses scarce resources with "no guarantee that the resulting social learning will yield a clear consensus at the end" (Reich 1990, 9).[1] But he believes that over time, it can be far more effective in helping to define and sustain mandates than either the traditional emphasis on efficiency or the administrative advocacy orientation has proven to be. Although deliberation takes time — and thus money — it can cost far less than many failed projects that have never received the public support necessary for effective implementation (Yankelovich 1999). Moreover, deliberation "can strengthen democratic institutions and the civic virtues on which such institutions ultimately depend" (Reich 1990, 8). In the view of Reich and his associates, deliberation is the solution to a powerful public sector left adrift without democratic guidance.

Civic Discovery in Environmental Protection

In their seminal analysis of the EPA, Landy and his colleagues (1994) find the root problem of the environmental agency to fall beyond the usual concerns with organizational maladies or ineffective public officials. Rather, the agency's failure can be traced more fundamentally to the system of interest group liberalism and its emphasis on rights-based policies that have shaped the ways public managers think and behave. Having fully infiltrated environmental policy making, interest group politics has turned the administrators into advocates of particular interests, a process Landy and his colleagues refer to as "policy entrepreneurship."

With the administrative environment thus transformed into a political arena, agency policy making has become a political competition among administrators and interest groups (Landy 1995). With administrators increasingly fashioning their own goals and programs, they find it necessary to contrive strategies to get around the various institutional and political impediments that stand in the path of carrying out their agendas. In efforts to dodge problematic legislators, judges, journalists, and interest group leaders, managers attempt to manipulate the system to achieve what they *see* as the most desirable policy. As Landy and his colleagues illustrate in their analyses of the Clean Air Act and Superfund (the major toxic waste clean-up law), effective administration ends up defined in terms of clever manipulation of the game.

But even when this approach works, it can pose serious problems for the environmental policy-making process more generally. As the circumvention of institutions and intermediaries gradually undermines government's credibility with the public, distrust and suspicion make it difficult to get things done in the future. The increasingly prevalent public view that administrators are making policy and manipulating the system to implement it erodes the overall decision-making capacity of government generally. Environmental officials are often rendered incapable of mustering support for the next round of decisions.

The problem is often worsened by agency professionals, as ecologists, economists, and lawyers tend to get caught up in their own versions of policy entrepreneurship. Each of these groups mainly sees its own professional orientation as more important than those of the others. Although each discipline clearly brings something important to environmental policy making, all of them necessarily offer only partial perspectives. In the competitive struggle to frame environmental problems, the real question gets lost: What is the relationship among these professional modes of inquiry? Given that none of them conveys all there is to know about a particular situation, how do we bring them together? Choosing among such alternatives requires pragmatic considerations. Which aspects of a situation are most important to explain? Does a given way of looking at a problem high-light critical choices by leaving out particular details? Is the situation obscured by oversimplifications? Does the neglect of these consider-ations leave out important options or consequences?

Advocacy groups, Landy et al. (1994) point out, take advantage of such disagreements to seek out and support experts and arguments that advance their own causes. In the process, expertise becomes the weapon of partisan conflict. The result is at best confusion about the appropriate role — or, perhaps better, limits — of technical knowledge, and at worst, the widespread belief that experts are mere hired guns who have nothing of great importance to contribute to the policy debate. In the process, both government officials and ordinary citizens find it difficult to learn about problems, as they do not know whose advice — if anyone's — to trust.

Following Reich's lead, Landy and his colleagues find the solution in reorienting environmental policy making and administration around the process of discourse. In their view, what is needed is an approach that brings together the full range of relevant participants in a more careful examination of the substantive merits of environmental issues. Indeed, as they see it, only through a process of deliberation can policy making be redirected away from an interest-driven policy entrepre-neurship toward the public interest more generally.

The mission of the EPA, they argue, should be to transcend interest group politics by provoking and leading comprehensive deliberations on pressing environmental issues. Such deliberations should involve all interested parties but must also include the broader public and its elected representatives. Fundamental to such leadership would be the ability to educate the public. EPA leaders, as facilitators of such delib eration, would limit their own role to the technical merits of a particu-lar problem. As professional experts, they would comment publicly on the technological, legal, and financial feasibility of a particular pro-posal. They should provide a scrupulous overview of what is known and what is not, illuminating both the empirical consequences and the ethical import of alternative policy choices (Landy 1995).

An agency such as the EPA has many opportunities to engage in such public education. In addition to testifying before Congress, EPA offi-cials issue a myriad of documents that appear in the Federal Register, prepare briefs for the president and other executive officials, hold press conferences, and make speeches. Each time the agency is obliged to respond to interest group comments, or to hold a public hearing regarding a proposed standard or regulation, it has the chance to explain and clarify the core issues raised by a proposed course of action.

As difficult as such changes might be to bring about, Landy and his colleagues argue that they would be well received by many segments of the EPA's constituency. In their view, members of Congress who view themselves as beholden to neither the environmental movement nor industry would benefit from a disinterested source of information. Similarly, those in the media who seek to report impartially on environmental affairs could use such information to better clarify key controversies. Moreover, a great deal of tension within the bureaucracy itself would be relieved, as many bureaucrats remain uncomfortable with policy advocacy and would welcome the opportunity to again become civil servants. Such administrators and their experts would be happy to trade the function of policy promoter or salesperson for one that is both more public spirited and more intellectually demanding. Such a role, Landy and his colleagues argue, would be of great assistance to top-level policy makers in desperate need of dispassionate advice.

Civic Discovery

Moral Virtue versus Democracy

Civic discovery has merits and provides a useful starting point for public sector reform. Missing from this approach, however, is a concern for a more active and engaged democracy. As formulated by Reich and his associates, top-level public managers can still remain part of a relatively elite policy community largely removed from the citizenry. While the approach would make available better information and arguments for ongoing processes of policy formulation, it remains unclear as to how it would bring citizens closer to these processes. The model seems to leave us more or less dependent on the virtues of the new public manager to educate and guide us to a democratic consensus (Roberts 1995). Although commendable, there is little about education in a political world that is apolitical. Moral virtue alone has seldom proven to be a reliable form of protection in such a world. In practice, morality never stands outside the social and organizational constraints that surround it. Unless the proposed deliberative process more clearly addresses the political relationship between public administrators and the public, the approach will remain top-down and elitist.

The alternative would be a more participatory relationship between the public facilitator and the citizenry. Toward this end, the approach could be democratically restructured by defining the role of the public manager more specifically as that of facilitator of *citizens'* deliberation. Rather than merely promoting public discourse for a mass society governed by a small number of elites, the goal could be to facilitate genuine citizen involvement in the pressing issues of the day. Basically, such a model would involve a form of participatory inquiry. Instead of government administrators directing the discussion, they would be responsible for setting up or structuring public contexts that help citizens organize their own discourses. While this might strike some as idealistic, we are not without such models.

Perhaps the classical example remains that of the Berger Commission in the 1970s in Canada. In this case, a judge appointed to assess the environmental impact of an oil pipeline designed to run through the traditional lands of northern Native Americans innovated a series of deliberative forums that brought together members of the local indigenous communities, industry representatives, and government and political officials. Heralded in many quarters as a major contribution to participatory policy inquiry and decision making, the reports of the commission were instrumental in halting a project widely viewed as portending social and environmental disaster for these Native American communities of northwestern Canada.

The Berger Inquiry

The Berger Commission inquiry was one of the oldest and most important experiments in participatory policy inquiry at the national level. In many ways the grandfather of such efforts, the Berger inquiry supplied the beginnings of a publicly oriented mode of participatory inquiry. It emerged as a response to political conflict over the Canadian Settlement Act of 1971 and the building of an oil pipeline through regions long inhabited by northern Native Americans, in particular the Inuit. To deal with the issue, a nongovernmental organization, the Alaskan Native Review Commission, was established by the Inuit Circumpolar Conference, a meeting of Inuit from Canada, Alaska, and Greenland. A Canadian judge, Thomas Berger (1977), was invited to conduct the commission investigations.

From the outset, Berger introduced a boldly innovative strategy for conducting the inquiry.[2] Rather than simply hiring experts to look into the questions of contention, he and his assistants went to some sixty rural fishing villages and camps throughout the settlement area. These were often distant and underdeveloped, and reaching them was often a feat unto itself. The objective was to offer each native Alaskan an opportunity to participate personally in the commission's inquiry. Moreover, the goal was not just to take testimony but rather to supply communities with information and education about the Settlement Act. Toward this end, Berger's staff provided workshops to acquaint the residents with both the key questions and the kinds of materials necessary to make informed decisions about the future of their region. In particular, the staff sought to draw out the communities' local knowledge concerning ways of life, economic and social needs, unique or peculiar local circumstances, and social values and beliefs. During the process, Berger acted as facilitator, often intervening with specific questions about conditions and concerns. In all, the commission heard from about 1,500 residences belonging to a total population of about 50,000 or 60,000 (Berger 1985).

What Berger did was to create a public space for groups otherwise largely cut off from, and ignored by, the rest of Canadian society. The commission provided a forum in which indigenous people could express, develop, and share their ideas and views. Existing outside the formal institutions of state and federal governments, the commission mainly received its funds from churches, foundations, and Native regional corporations. As such, it carried out its mission without either an endorsement or monies from official Canadian bodies.

At the conclusion of the inquiry, the commission offered a report that challenged federal and state authorities. Rather than ratifying the construction of the government's oil pipeline with the argument that it would bring new employment possibilities for the Native American communities, the recommendations were based largely on Native concerns. They called for strengthening the subsistence economy of the area and rejuvenating tribal governments. Not only did the report offer a picture of the community, but it spoke to the needs of the citizens as they understood them.

Later, Berger used his model in other royal commissions over which he presided. These included issues of family and children's law in British Columbia, the construction of oil pipelines from the Arctic to southern markets, and Inuit and Indian health care. Of all the efforts, the Mackenzie Valley Pipeline Inquiry was the most important success (Berger 1977). It served to help persuade Canada's federal government to strengthen the renewable resources and Native settlement claims of northern communities before building any new oil pipelines.

The Alaska Native Review Commission was thus an unprecedented innovation in policy inquiry. It can be judged as doing nothing less than assisting the Inuit peoples to form a political community that permitted them to think about their future. In political scientist John Dryzek's (1990, 127–28) words, "It provided a forum for conflict resolution within that community through the transformation of private into public norms through sustained discussion." As Dryzek maintains, the practices of the commission provide a model for participatory policy science.

Berger, in this respect, was a true pioneer. Much of his method was invented as he went along. There was little alternative; he had no precedents to follow. Since then, however, there have been a number of attempts to systematically develop procedures for such forums. None have been more important or impressive than the Danish consensus conference.

Since the days of the Berger Commission, there has been a growing body of research focused on how to bring citizens' preferences to bear on public policy (Lindeman 1997; Fishkin 1996; Bohman 1996; London 1995). Moreover, there has been a wide range of projects and experiments with such practices, especially in northern Europe. These include "citizens panels," *"Buergergutachten"* (citizen evaluations), "scenario workshops," and consensus conferences, among others (see appendix F).

All these forms bring citizens together to assess complex environmental and technological issues. Experts are present only to supply information and answer questions as the citizens find necessary. Although little known to either academics or the general public, these experiences offer important insights about how to bring citizens closer to public decisions processes. Most important, they have shown that citizens are capable of much more involvement in technical questions than is conventionally presumed.

In the next section, we present the Danish model of the consensus conference, the most sophisticated of such participatory practices. Whereas most methods for citizen deliberation have dealt with rather narrowly defined issues at the local level, the Danish consensus conference is institutionalized at the national level and addresses broad social and economic questions. Moreover, while most forms of citizen panels operate behind closed doors, the consensus conference is open to the public as a whole. Indeed, one of its primary purposes is to inform and stimulate broad public debate on the given topic.

The Consensus Conference

The consensus conference, developed by the Danish Board of Technology, has emerged as the most elaborate form of citizens' panel.[3] Inspired in the 1980s by the now defunct U.S. Office of Technology Assessment (OTA), the Board of Technology sought a new and innovative way to get around the divisive conflicts associated with environmentally risky technologies such as nuclear power. The idea was to find a way to make good on OTA's original mission of integrating expertise with a wide range of social, economic, and political perspectives (Joss and Durant 1995). Toward this end, the board developed a model for a "citizen's tribunal" designed to stimulate broad social debate on issues relevant to parliamentary level of policy making. In an effort to bring lay voices into technological and environmental inquiries, the board sought to move beyond using narrow expert advisory reports to Parliament by taking issues directly to the public. In compliance with the long-standing Danish political tradition of "people's enlightenment," which stresses the relationship between democracy and a well-educated citizenry, the board developed a framework that bridges the gap between scientific experts, politicians, and citizens (Kluver 1995; Mayer 1997).

The formal goals of the consensus conference are twofold: to provide members of parliament and other decision makers with the information resulting from the conference, and to stimulate public discussion through media coverage of both the conference and the follow-up debates. First implemented in 1987, conferences have dealt with issues such as energy policy, air pollution, sustainable agriculture, food irradiation, risky chemicals in the environment, the future of private transport, gene therapy, and the cloning of animals.

A consensus conference is organized and administered by a steering committee appointed by the Board of Technology. Typically, the conference involves bringing together ten to twenty-five citizens charged with the task of assessing a socially sensitive topic of science and technology. The lay participants are usually selected from written replies to advertisements announced in national newspapers and radio broadcasts. Interested citizens, *excluding* experts on the particular theme, apply by sending a short written statement explaining why they would like to participate in the inquiry process. The steering committee evaluates the statements to determine if a candidate is sufficiently dedicated to participate fully in the conference process. Citizens participate as unpaid volunteers, but the board offers compensation for any loss of income that might occur as a result of the involvement. From around 200 to 300 written responses, the board selects an average of fifteen citizens. Although the groups do not constitute a random sample of the population, they are selected on the basis of sociodemographic criteria such as education, gender, age, occupation, and area of residence. As such, a panel is generally a reasonable cross section of ordinary citizens with no special interest or knowledge in the topic under investigation.[4]

From choosing a topic to the final public discussions after the formal conference proceedings, the process typically runs several months. Central to the inquiry process is a facilitator who assists the lay panel in completing its tasks. Professionally trained in communication skills and cooperative techniques, the facilitator is a nonexpert in the topic of the conference. Working closely with the panel, he or she guides the process through an organized set of rules and procedures. In addition to organizing the preparatory informational and deliberative processes, he or she chairs the conference. Somewhat like a judge in a jury trial, the facilitator maintains the focus of the experts on the lay panel's questions during the conference and assists panelists in finding the most direct answers to their questions.

After the lay participants are selected, they assemble for informal meetings on the topic. At the first meeting, the steering committee outlines the topic for the participants in general terms and informs them that they may define their approach to the topic in whatever way they see fit. Not only can they frame their own questions, but they can seek the kinds of information they find necessary to answer them. At the same time, panelists are supplied with extensive reading materials by the steering committee and given a substantial interval of time to study the materials at home.

After reading the materials, the panelists are asked to develop a list of questions pertinent to the inquiry. The steering committee uses the participants' questions to assemble additional information for the panelists and to identify an interdisciplinary group of technological and environmental experts to make presentations to the citizen panel. During a subsequent informational meeting, the citizens review new materials and further refine their list of questions, dropping some as well as adding new ones. Evaluations of this phase show that by the time of the actual conference, the participants are remarkably knowledgeable about the issues at hand. In some cases, a hearing is also organized for parties interested in the selected subject. Such groups — for example, individuals or companies with extensive knowledge, influence, and dependence on the field; research institutions; research committees; traditional interest groups; and grassroots organizations — are provided with an opportunity to contribute information to the deliberative process. These hearings may either be in writing or organized as meetings. The information culled is used to further both the organizational work of the steering committee and the thought and discussion of the lay panel.

The official conference begins at this point and typically lasts three or four days. On the first day, the experts make presentations running twenty to thirty minutes. After each presentation, the members of the lay group question the experts and cross-examine them. In some cases, representatives from relevant interest groups are present and can also be questioned. Within the given time limits, citizens in the audience are also invited to make statements or ask questions. On the second day, the citizen panel more actively cross-examines the experts. Again, at specific points, the public and interested parties are encouraged to ask questions. In some cases, representatives from relevant interest groups are also questioned.

At the end of this process, usually on the third day of the meeting, the citizen panel retires to deliberate on the exchanges. With the assistance of a secretary supplied by the steering committee, the group prepares a consensus report (fifteen to thirty pages in average length) that considers all of the issues that bear on the topic. Typically, the report reflects the range of interests and concerns of parties involved in the conference. Beyond scientific and technical considerations, the report speaks to the spectrum of economic, legal, ethical, and social aspects associated with the topic.

On completing the report, the citizen panel publicly presents its conclusions. Normally this takes place in a highly visible public setting in the presence of the media, a variety of experts, and the general public. Subsequently copies of the report are sent to members of Parliament, scientists, special interest groups, and members of the public. Consensus conference reports often complement expert assessments as part of larger technology assessment projects.

Described as an exercise in "counter-technocracy," the consensus conference has received favorable reviews from citizens, experts, the media, and politicians. Many Danish politicians have responded particularly favorably to the approach. Because they are themselves laypersons, they can easily identify with the inquiry process and its outcomes. They also find that the conclusions more clearly reflect the concerns of the population than do the more traditional expert assessments. An indication of this favorable evaluation is found in the positive impacts consensus conferences have had on parliamentary decisions in a range of topics pertinent to Danish environmental protection (Kluver 1995, 44). For example, panel recommendations have influenced the Parliament to decide against funding animal gene technology research and development programs, to restrict food irradiation, and to accept a panel proposal for a tax on private vehicles.

No less significant is the impact the consensus conference experience has had on the citizen participants. Joss (1995, 3) reports that lay panelists report both an increased knowledge of the subject under discussion and a new confidence in their ability to deal with technical interests generally. Equally important, panelists tend to describe the conference experience as having supplied "a stimulating and creative input to their personal life." Joss quotes one participant as having said that the experience provided her with "an increased interest in all sorts of subjects" that she had previously thought were "over my head," as well as the discovery that she "could actually understand (and comment on) scientific issues."

Since the outset of the Danish experience, consensus conferences have spread to Austria, Holland, Britain, New Zealand, Norway, and Switzerland (Joss 1998). A preliminary pilot project dealing with the topic of telecommunications was also conducted in Boston in 1997 by the Loka Institute. As a highly innovative contribution to the facilitation of democratic practices, the consensus conference provides a model for giving citizens a role in the environmental policy process. Not only has it been widely credited with invigorating contemporary democratic practices, but it has also built understanding and trust among citizens and experts. As such, the consensus conference would appear to provide a model for a democratization of the civic discovery process that Reich and his colleagues promote.

The Consensus Conference as Civic Discovery

The consensus conference represents an important innovation for participatory inquiry, in particular for its institutionalization in a modern political system. It shows both that citizens are capable of dealing with complex technical questions and how a citizen's panel can be organized to feed into established political decision processes. It points, moreover, to ways in which professional advice can be reoriented to serve a deliberative process of policy inquiry. The consensus conference has thus not only earned the praise it has received in countries where it has been put into practice but also deserves wider political attention in general, especially in the academic field of policy analysis.

As a method for civic discovery, understood as participatory democracy, only two dimensions of the consensus conference process need further attention. One concerns time. In most consensus conferences, citizens have only about a half day at the end of the proceedings to discuss and deliberate on their findings. If the goal is to move citizen panels beyond problem solving to democratic participation more generally, such a time period is insufficient for normative deliberation. Indeed, many conference panelists have registered this point. Reflexive consideration of basic social assumptions is not something that can be dealt with summarily. It requires an iterative process of consideration and reconsideration of different perspectives and ideas.

The time limits typically allowed for deliberation are not altogether the result of pressures imposed by cost or related factors. Indeed, Danish organizers have argued that if experts can summarize their complex analyses in a half-hour presentation, citizens should be able to bring their own value orientations to bear on the deliberations within a similar time frame (Kluver 1995). This argument, however, rests on a false assumption about normative deliberation. It assumes, as does much of contemporary social science, that attitudes and beliefs simply exist and only need be identified and portrayed. This view misses the point of deliberation. Beyond merely uncovering normative assumptions and beliefs, deliberation can lead to changes in assumptions, as well as the creation of new ones. To discourse about policy values and norms is to reflect as much on what they *should* be as what they *are*. While determining what citizens believe at any particular time is useful for political decision makers, it does not serve the requisites of a fully developed democracy. Not only is democratic participation a method for deciding what ought to be, but democracy is a means for advancing the human community to a higher level of social awareness and understanding. Beyond merely supplying political decision makers with information about what citizens think about a particular topic, democracy offers citizens the unusual opportunity to deliberate publicly on the crucial political and social questions of the day. As such, it provides an important vehicle for revitalizing the public sphere.

Experience with the consensus conference shows that participants are able to discuss the issues on this level.[5] Without assistance, many participants have proven able to raise basic social and philosophical issues. For example, in a consensus conference concerned with the genetic altering of plants, participants discussed not only the global implications of such research for agriculture in Europe and the Third World but the meaning for humanity itself. The problem is only that the time generally allotted for deliberation militates against much more than the mere mention of such issues. Ameliorating this problem poses no major challenge. Adding a day or two, if not a week, to the process is in no way insurmountable.

More challenging is a second problem that needs to be addressed, namely, the relation of the citizens to science and expertise. Examination of citizen deliberations in consensus conferences shows that there is little opportunity to question or examine science itself. This raises an especially important issue for normative deliberation on complex technical questions. We have already seen that crucial normative assumptions are often buried in the technical analysis itself. Who, for example, can know that particular assumptions and value judgments were in operation without looking more closely into the research itself? Such investigation, of course, is a different order of challenge that requires not only more time but a particular kind of guidance, if not training. It would be the sort of thing better suited for participatory research, which often brings the citizens themselves into the research process.

Conclusion

Participatory inquiry need not be limited to small groups concerned with local issues. The Danish Consensus Conference illustrates that citizen deliberation on environmental and technological issues can play an important role in forming national public opinion with regard to significant legislative policy questions. As such, it offers a unique way to invigorate democracy, as well as to restore understanding and trust between experts and citizens.

The model also offers insights into ways to strengthen the concept of civic discovery. As the theory has emerged, it relates to public managers as individuals, saying little or nothing about the structures that would make possible such a deliberative approach. Rather than leaving the discursive process to an administrative elite with a high-level civic conscience, participatory inquiry and the Danish model in particular suggest how a discursive model of public administration can be built into the institutional policy-making structures as well. Without such structures, the theory of civic discovery runs the risk of producing only a new administrative class in the name of democracy. By distributing the emphasis across citizens, experts, and policy makers, the consensus conference transcends this difficulty.

Finally, such innovations in participatory inquiry pose important, challenging questions for future research. In particular, what can they tell us about the interactions between citizen and expert knowledges? What can we learn from such innovations about the interactions of technical and cultural factors in such assessments? What, more specifically, is the role of the citizen participant's local knowledge in assessing truth claims of opposing scientific experts? With these questions in mind, we turn in the next and final chapter to the "epistemics" of such deliberative practices.

12. The Environments of Argument

Deliberative Practices and Policy Epistemics

The classical models of reasoning provide inadequate and in
fact seriously misleading accounts of most practical and academic
reasoning—the reasoning of the kitchen, surgery and workshop,
the law courts, paddock, office and battle field; and of the discipline. . . . Common reasoning often has components that are or can
usefully be represented as tidy demonstrations governed by the
logics of deduction or probability; the problem is that they are only
components, and a completely distorted picture of the nature of
reasoning results from supposing that these neat pieces are
what reasoning . . . is all about.—Michael Scriven

Throughout the preceding chapters, I have emphasized the importance of citizen participation in complex environmental decisions.
While we have learned that there are no easy answers, the discussion
has shown the conventional wisdom to be overly simplistic. The argument that citizens are unable to participate in making technical policy
decisions goes against the credible realities of important experiments.
As the practices of popular epistemology, participatory resource mapping, and the consensus conference have shown, many ordinary citizens are quite capable of grappling successfully with both the technical and the normative issues that bear on environmental decision
making.

    This has not been to argue that scientific expertise is unimportant.

Given the complex nature of environmental problems, there can be no escape from the need for scientific expertise. Professional expertise will remain basic to the discovery of the environmental problems and will continue, as well, to play a central role in innovating policy solutions. For this reason, I have argued that a democratic society needs to significantly rethink the role of policy expertise vis-à-vis the citizen. Toward this end, the experiences of participatory inquiry have been shown to offer important sources of practical insights. Not only does participatory inquiry represent an important contribution to democratic theory and practice, but it contributes to expertise itself. Beyond supplying the legitimacy necessary for democratic decision making, citizens can bring important contextual information about local characteristics and related circumstances to bear on policy problem solving and decision making. Policy experts, in this respect, are at times as much in need of the citizens as the citizens are of the experts. In addition to the "citizen expert," this suggests a rethinking of the expert's role as that of "specialized citizen."

Without denigrating environmental science, a frequent practice among radical environmentalists, the discussion has emphasized the limits of the field's policy-relevant knowledge. The experience with environmental science in policy making as we have seen, has been defined more by uncertainties than by clear answers. The largely unintended result has been a turn away from a strictly technocratic approach to a politics of expertise, featuring experts and counterexperts. Although this has represented an improvement over the technocratic orientation, it has also had unfortunate consequences. It has, in short, offered the opponents of environmental protection a way to counter policies they don't like: they simply hire experts who share — or are willing to represent — the views they want to hear. The result has been a paradox. As Beck explained, we are dependent on a method that cannot answer our questions. This dilemma, as we saw in chapter 7, has led citizens involved with environmental decisions to emphasize cultural forms of reason.

Citizen participation in policy inquiry contributes to three important goals. First of all, it gives meaning to the practice of democracy. All too often today, democracy is only given lip service. If we are to take the idea of a strong democracy seriously, as Barber (1984) puts it, all citizens need to participate at least some of the time in the decisions that affect their lives. Participatory inquiry, as we have seen, can play an important role in helping to bring that about.

Second, citizen participation in the policy inquiry process can contribute normatively to the legitimization of policy development and implementation. Participation, in this respect, can be understood as helping to build and preserve present and future decision-making capacities (Diesing 1962, 169–234). Discursive participation, in this sense, offers the possibility of getting around the debilitating effects of interest group competition inherent to liberal pluralism (R. Hiskes 1998). Taking aim at the competitive interest group bargaining at the root of many policy failures, in particular environmental policy failures, the collaborative consensus building inherent to participatory discourse makes possible the identification and development of new shared ideas for coordinating the actions of otherwise competing agents (Busenberg 1999). Or as Healey (1997, 30) explains it, by transforming ways of organizing and knowing, such collaborative deliberation has the possibility of building new political cultures.

And third, but not least important, we have shown that citizen participation can contribute to science itself. Participatory research, in its various forms, has the possibility of bringing to the fore new knowledge — in particular local knowledge — that is unavailable to more abstract empirical methods. Indeed, the ability to deliver first-hand knowledge of the circumstances of local context is seen to address a major limitation of conventional methods. This, in fact, led us to examine local knowledge in the light of a postpositivist understanding of science, in particular an approach capable of integrating the general and the particular in the context of a larger deliberative perspective.

The contributions of participation, as we have seen, are central to the pursuit of environmental sustainability. Research shows that participation has played an important role in environmental change generally. Jaenicke (1996), for example, demonstrates that all significant environmental efforts have had citizens' efforts behind them. Moreover, importance of participation now receives official acknowledgment in important environmental policy documents. For example, the Brundtland report, prepared for the Rio Summit in 1992, states that success in achieving sustainability "will depend on the widespread support and involvement of an informed public," calling for an expanded role of their "participation in development planning, decision-making, and project implementation." Recognizing that effective policies have to be built on evolving patterns of everyday life, the Rio Summit spelled out a program for promoting sustainable development at the community level of the citizen — namely, the Local Agenda 21 Action Program. Given that 60 percent or more of the chores associated with environmental sustainability are anchored to local concerns, the program has an important contribution to make to the success of Agenda 21 as a whole (Selman 1997, 21). Unfortunately, post-Rio analyses of Agenda 21 show this local dimension of the plan to be either weakly implemented or neglected. As Irwin (1995, 135) puts it, the public is still "seated ringside but certainly not at the centre of the environmental action — at least so far as 'official' decision-making processes are concerned."

This is not to overlook the active citizen-oriented interest groups that have had a significant impact on environmental policy. Indeed, environmental protection, as we saw, has in significant part emerged through interest group activities and, as such, has given rise to mainstream environmentalism. Without denying the importance of interest groups, however, they should not be confused with citizens. Although interest groups represent citizens, especially "public interest groups," they are hierarchical organizations often rather distantly removed from the citizens for whom they speak. In fact, over time, interest groups tend to become part of the configuration of governing interests and at times come to be seen as part of the problem. The environmental groups have been no exception. For this reason, I have argued, interest groups need to better avail themselves of the citizens' views that the groups seek to represent.

Toward a Deliberative Policy Analysis

Beyond a general call for citizen involvement, environmental interest in participation has also begun to take more concrete forms. Of particular interest, in this respect, is the report of the National Research Council, *Understanding Risk: Informing Decisions in a Democratic*

Society. The Council, an arm of the U.S. National Academy of Sciences, not only calls for environmental participation but takes the next step as well, advocating a citizen-oriented deliberative approach to risk analysis. For more than fifteen years, the Council's books on risk assessment and risk management have served as a principal source of information and guidance on risk assessment and management. The series as a whole can be read as a chronicle of a scientific confrontation with a particular sociotechnical reality, namely, the complicated question of describing and measuring risky technologies. The most striking feature about the chronicle is that it is, albeit unintendedly, a story of a struggle with the limitations of neopositivist science in the world of practical affairs.

Understanding Risk makes clear the degree to which the Council's views have evolved. In this work, the Council calls for an "analytic-deliberative approach" to environmental and technological risk analysis. Against the Council's early technocratic writings in the 1980s, this book can only be seen as a major advance. Not only does the Council now acknowledge the need to bring local participants into the process, but it also calls for a deliberative method of inquiry. While in the early 1980s citizen participation was seen more as the problem than the solution, it is now judged to be an essential requirement, especially in situations that combine high degrees of uncertainty with low levels of trust.

In *Understanding Risk*, coping with risky situations is seen "to require a *broad understanding* of the relevant losses, harms, or consequences to the interested and affected parties, including what the affected parties believe the risks to be in particular situations" (National Research Council 1996, 2). For this reason, it becomes necessary to incorporate their perspectives and specialized knowledges into risk decision making. Put in different terms, the citizens' local knowledge must be incorporated into risk assessment.

The report counsels organizations to make special efforts to ensure that the interested and affected parties find reasonable basic analytic assumptions about risk-generating processes and risk estimate methods. Even though potentially more cumbersome and time-consuming in the short run, as the study asserts, it is generally better to err on the side of too-much rather than too-little participation. Organizations are advised to rigorously evaluate the need for involvement of the full range of affected and interested groups at each step in the process. This is all the more necessary when the stakes in a decision are high and the public's level of trust in the responsible organization is low. Under these circumstances, adequate risk analysis "depends on incorporating the perspectives and knowledge of the interested and affected parties in the earliest phases of the effort to understand the risks" (3). The challenges "of asking the right questions, making the appropriate assumptions, and finding the right ways to summarize information can only be met by designing processes that pay appropriate attention to each of these judgments, inform them with the best available knowledge and the perspectives of the spectrum of decision participants" (3). Only through such processes can organizations make choices and decisions that affected groups will accept and trust.

Basic to this challenge is the need for "an analytic-deliberative method" capable of bringing together citizens and experts. Such a deliberative method is required to guide a participatory process that can "broadly formulate the decision problem, guide analysis to improve . . . participants' understanding [of decisions], seek the meaning of analytic findings and uncertainties, and improve the ability of interested and affected parties to participate effectively in the risk decision process" (3). The process must have an appropriately diverse participation or representation of the spectrum of interested and affected parties, of decision makers, and of specialists in risk analysis at each stage of the process. Most important is the need for participation in the early stages of problem formulation.

In this view, analysis and deliberation are presented as complementary approaches to gaining knowledge about the world, forming understanding on the basis of knowledge, and reaching agreement among participants. Whereas analysis "uses rigorous, replicable methods, evaluated under the agreed protocols of an expert community," deliberation is a process in which participants "discuss, ponder, exchange observations and views, reflect and attempt to persuade each other" (3–4). Deliberation, moreover, does not come only at the end of the process. It is important at each step of the process that informs risk decisions, such as deciding which harms to analyze and how to describe scientific uncertainty and disagreement. Appropriately structured, deliberation contributes to sound analysis by adding knowledge and perspectives that improve understanding, and contributes to the acceptability of risk characterization by addressing potentially sensitive procedural concerns. As the study sums it up, "deliberation frames analysis, analysis informs deliberation, and the process benefits from the feedback from the two" (6).

In a discussion of quantitative analysis and its effort to reduce the many dimensions of risk to a single metric, as in the cases of risk- and cost-benefit analysis, the study stresses the difficulties resulting from leaving out the normative dimensions of such judgments. Because such methods necessarily simplify real-world situations, as the study puts it, value judgments and other normative assumptions remain implicit and overlooked. For this reason, successful use of such techniques depends on their interaction with deliberative methods that help to bring out these normative dimensions in the very design of the analysis, including the determination of the particular methods to be employed and the interpretation of findings.

Although the process of deliberation needs to be broader and more extensive for some cases than others, explains the Council, such discourse is now deemed basic to all risk assessments. For organizing such a deliberative process, the Council suggests a number of guidelines. Perhaps most interesting is the view that organizations should take the extra initiative to reach out with technical assistance to unorganized or inexperienced groups in matters of risk analysis and regulatory policy.[1] "If some parties that are unorganized, inexperienced in regulatory policy, or unfamiliar with risk-related science are particularly at risk and may have critical information about the risk situation, it is worthwhile for responsible organizations to arrange for technical assistance to be provided to them from sources that they trust" (4). Without using the language of participatory inquiry, the Council suggests that experts must at times assume the role of facilitators. It also suggests that greater attention be given to discursive strategies such as deliberative polling and citizens' juries, although the book mainly relegated this discussion to an appendix.

Moreover, the convenor of such analytic-deliberative processes should "clearly and explicitly inform participants at the outset about the legal, budgetary, or other external constraints likely to affect the extent of deliberation . . . or how the input from deliberation will be used" (5). The deliberative process, in this respect, strives "for fairness in selecting participants and in providing, as appropriate, access to expertise, information, and other resources for parties that normally lack these resources" (5). Last but not least, "managers should build flexibility into deliberative processes, including procedures for responding to requests to reconsider past decisions or to change procedures within externally established limits of time or resources" (5).

Deliberation, in this perspective, cannot be expected to end all controversy. "It will not guarantee that decision makers will pay attention to deliberation's outcomes, prevent dissatisfied parties from seeking to delay or override the process, or redress the situation in which legal guidelines mandate that decisions be based on a different set of considerations from those that participants believe appropriate" (5). Controversies, in this respect, are constructive in helping to identify weak points from which science can profit rather than merely as barriers in the path of expert decision making. Not only do controversies encourage in-depth analysis to identify and explicate a technology's risks and benefits and their social implications, but they can provide partly conflicting assessments of new technologies or of the environmental impacts of actual or proposed projects that are further articulated and consolidated in the course of a controversy. In this view, the proper function of a controversy is to identify and evaluate potential problems. The analysis of them serves, as such, as an informal complement to conventional methods of risk assessment.

Against these technocratic beginnings in risk analysis, to which the Council has made some of the most important contributions, the report can only be judged as an impressive advance in the thinking of the Council. Although technocratic practices still remain dominant, the fact that many of the contemporary methods were initially influenced by the Council holds out some hope for change, even if slow and reluctant. The recognition of the centrality of participation and discourse by this prestigious body should not be underestimated.

Having acknowledged the Council's latest contribution as a step forward, however, it is important to point out its shortcomings. In the context of a postpositivist perspective, the Council still pulls up short of fully recognizing the implications of its own position. In this respect, the Council advances its position conservatively, often underplaying or ignoring important implications of its analytic-deliberative recommendations. First of all, deliberation is mainly relegated to the task of supplying information to those in the business of risk analysis. And risk analysis itself remains a scientific, expert-driven undertaking. In this respect, the Council is not ready to suggest that there might be a sharing of the process. Experts should deal with factual disputes; participation is restricted to the normative dimensions.

The argument that discourse and analysis complement each other stops short of a postpositivist perspective. As the Council understands this relationship, science can still stand alone analytically. What it fails to take account of is the degree to which science cannot exist independently of normative constructions. While the fact that the meanings of such findings are always interpreted in a particular social context is to some extent recognized — whether in the context of an expert community or society more generally — the Council neglects the deeper realization that such research is itself built on normative assumptions. That is, it is embedded in the very understandings of the objects and relationships that are under investigation. In this sense, the goal of discourse is not just "to improve understanding" but rather to *create* it through exploring the social meaning of the research and its findings.

The Council's model thus still remains attached to the conventional notions of science, in particular the division between the empirical and the normative dimensions. In short, the Council never opens up science to critical scrutiny. It acknowledges all of the conditions for a deliberative approach but never recognizes or concedes that such a revision also has profound meaning for science. Deliberative participation, as it is advanced in this work, remains outside of science. What is needed is an effort to move away from the idea that the two models complement each other — that is, as two separate things that can inform each other — and to see them as a continuation of the *same* activity. As deliberation occurs in both normative and empirical research, they need to be understood as two dimensions on the same spectrum. Of particular importance is the fact that normative elements lodged in the construction of empirical research need to be accessible for deliberation. Indeed, the very construction of the empirical object is sometimes at stake. So too are a wide range of assumptions that empirical researchers make as they apply their analytic tools to a given reality.

Facilitating Policy Deliberation

Communicative Interaction and Postpositivist Inquiry

The environmental crisis, as we have seen, has played a central role in compelling us to rethink scientific expertise and its relations to public decision making. In this regard, I have examined the concept of a postpositivist science and its implications for citizen participation. Postpositivism makes clear the importance of participation in determining the relation of social meaning to empirical information. Furthermore, participation can also confer legitimacy on analysis, of special importance in environmental analysis given the almost inevitable presence of uncertainty and debate.

For expertise, participation means reconstructing the activities of the analyst to emphasize the facilitation of deliberation (J. Forester 1999). As we discussed in chapter 9, this involves rethinking the relationships among analysts, citizens, and decision makers. Establishing an open and democratic exchange requires bringing these roles together in mutual exploration. Methodologically, an approach capable of facilitating the kind of open discussion essential to a participatory context is needed. Such a method would provide a format and a set of procedures for organizing the interactions between policy experts and the lay citizens they seek to assist. As we saw, this involves establishing a collaborative or participatory relationship with the citizen-client (Hawkesworth 1988; Schon 1983; Healey 1997). Albeit in quite different ways, writers such as deLeon (1992), Durning (1993), Fischer (1990), and Laird (1993) have called for such a "participatory policy analysis."

In the approach to participatory inquiry outlined in chapter 9, the expert serves as "facilitator" of public learning and political empowerment. Rather than to provide technical answers designed to bring political discussions to an end, the task of the analyst-as-facilitator is to assist citizens in their efforts to examine their own interests and to make their own decisions (Fischer 1992). Although this conception of the expert's role differs sharply from the standard understanding, it is not as new to the policy literature as it might seem. Indeed, Harold Lasswell (1941, 89), the founder of the policy science movement, first envisioned the role of the social scientific policy professional as that of "clarifier" of issues for public deliberation. Following Dewey's call for improving the methods and conditions of public debate, Lasswell defined the professional's role as that of educating a citizenry capable of participating intelligently in deliberations on public affairs. Policy science was thus initially seen as a method for improving citizen deliberation. Toward this end, Lasswell spelled out a "contextual orientation" for professional-analytic practices that would extend policy science beyond the professional realm to include the insights and judgments of the citizenry.

If this contextual orientation was lost to the subsequent development of a technocratic policy science, it has more recently returned in the policy-oriented literatures of public administration and planning, especially in the postpositivist and postmodern literatures of these fields. In postmodern public administration theory, for example, the administrator seeks to facilitate and clarify communication rather than decide which group is right (Rosenau 1992, 86–87). In the process, as Caldwell (1975, 567) argues, the administrator must aim for "foresight, initiative, flexibility, sensitivity and new forms of knowledge" that are not truth claims, or technological or procedural, so much as "interactive and synergistic." By the 1990s, such ideas had moved to the forefront of theoretical discussions in many circles in the fields of public administration and planning. Most of the advocates of this view are quick to argue that the approach is neither nihilistic nor hostile to reason. Rather, it is a search for new forms of knowledge and reason that carry us forward without the pretense of an unmovable universal "Truth." Basic to the effort is an emphasis on local knowledge.

Resting on a postpositivist epistemology, this "argumentative" or "communicative" turn in planning and policy analysis "takes as central the subjective notion of meaning and regards socially shared beliefs to be 'constitutive' of reality" (Innes 1990, 32). As we saw in the previous chapter, such a postpositivist analysis emphasizes a critical assessment of the assumptions that organize and interpret our ways of knowing and the knowledge that results from them. Such an interpretive approach is thus "distinctive in saying that knowledge is not the exclusive province of experts, and in accepting a subject element in all knowledge" (Innes 1990, 32). Rather than hide behind the guise of value neutrality, the expert must actively employ his or her own subjectivity to understand the views of others — citizens, politicians, and decision makers, among others. More than just an alternative epistemological orientation, it provides a more useful and realistic description of the actual relationship between the citizen and expert (Fischer and Forester 1993; Healey 1997; Innes 1990).

Grounded in particular social contexts, interpretive knowledge requires the professional to involve him- or herself in the modes of thought and learning of everyday life, that is, the local knowledge of the ordinary citizen. Knowledge is in this way recognized to be more than a set of relationships among selected data or variables isolated or abstracted from their social context (Innes 1998). To be meaningful for the world of decision and action, such variables have to be interpreted in the situational contexts to which they are to be applied. In this view, as we saw in chapter 4, what we call "knowledge" of the social world is the product of a negotiation between those with more "expert knowledge" and the participants in the everyday world, including the experts themselves. Moreover, as Innes (1990, 32) puts it, such "knowledge is about whole phenomenon rather than simply about relationships among selected variables or facts in isolation from their contexts." As such, knowledge and reasoning are recognized as taking many forms, from empirical analysis to expressive statements in words, sounds, and pictures (Healey 1997, 29). Of particular importance is the narrative form of the story; in everyday life, it is the primary means of giving meaning to complex social phenomena.

Bringing together professional knowledge and lived experience, citizens and experts form what might be thought of as an interpretive community. Through mutual discourse, this community seeks a persuasive understanding of the issues under investigation. This occurs through a transformation of individual beliefs, including social values. In this process, the inquirer, as part of a community, is an agent in the social context rather than an isolated, passive observer. Means and ends are inseparably linked in such a discursive process, and importantly, those who participate need to accept the practical and moral responsibilities for their decisions and their consequences.

The postpositivist facilitator also accepts the task of working to embed an inquiry in actual organizational and policy processes. This would include developing arenas and forums in which knowledge can be debated and interpreted in relation to the revelant policy issues.

One example of such an inquiry process is the consensus conference examined in chapter 11; another is the people's planning process in Kerala. Ideally, the goal is to establish institutional mechanisms for using the resulting knowledge, as illustrated by the poverty assessment discussed in chapter 9. This can involve incorporating findings in the regular work of an implementing agency through face-to-face communication among experts, citizens, and decision makers, as was the case in the Georgia state agency in chapter 8.

The interaction between the social scientist as facilitator and the citizen-clients can be likened to a conversation in which the horizons of both are extended through mutual dialogue. Building such a conversation, Jennings (1987) suggests, requires the analyst "to grasp the meaning or significance of contemporary problems as they are experienced, adapted to, and struggled against by the reasonable, purposive agents, who are members of the political community." He or she must then work "to clarify the meaning of those problems" in such a way "that strategically located political agents (public officials or policy makers) will be able to devise a set of efficacious and just solutions to them." Emphasizing a procedural route to policy choice, the analyst strives to interpret the public interest in a way that can survive an open and nondistorted process of deliberation and assessment. In the process, interpreting the world and changing it are complementary endeavors. The analyst-as-counselor seeks to "construct an interpretation of present political and social reality that serves not only the intellectual goal of explaining or comprehending that reality, but also the practical goal of enabling constructive action to move the community from a flawed present toward an improved future" (Jennings 1987, 127). Without necessarily buying into his methodology as a whole, we might liken the process to what Roe (1994) describes as developing a "metanarrative."

Policy Epistemics
Deliberation and the Fields of Argumentation

Given the central role of these socio-epistemological issues, coupled with their sophistication, there is a need in policy inquiry for a new underlying specialization that might suitably be called "policy epistemics." Borrowing from Willard (1996, 5), epistemics address "the predicaments of modern decision-makers," namely, their dependence on knowledge and authority, "their inability to assess the states of consensus in disciplines, their incompetence in the face of burgeoning literatures, and their proneness to mistaken agreements." Toward this end, policy epistemics would focus on the ways people communicate across differences, the flow and transformation of ideas across borders of different fields, how different professional groups and local communities see and inquire differently, and the ways in which differences become disputes.[2] Of particular importance in this respect is the interaction between expert inquiry and the processes of political and policy argumentation.

Willard proposes the "field of argument" as a unit of analysis. Focusing on how people construct their policy arguments, policy epistemics examines the interplay between specific statements or claims. By argument, he means polemical conversation, disagreement, or dispute, the principal medium by which people (citizens, scientists, and decision makers) maintain, relate, adapt, transform, and disregard contentions and background consensuses (on which conversations by particular groups are constructed — traditions, practices, ideas, and methods).

A field of argument, according to Willard, refers to a discursive terrain of inquiry organized around particular judgmental systems for deciding what counts as knowledge as well as the adjudication of competing claims. Such communities of inquiry, as vertically structured social entities, are defined as much by their disputes as by their agreements. Although fields vary in the degree to which they inspire confidence, most policy-oriented fields are sufficiently public and open to criticism that people have enough confidence in them to consider them justifiable. One reason for that confidence, according to Willard, is the belief that the field itself can be held accountable in terms of its grasp and reach. One can hope or expect to identify movement toward the achievement of the field's ideals, hopes, and ambitions.

For policy epistemics, this mean focusing on arguments and debates that constitute and shape the various policy networks or "policy communities" (for example, the network of social scientists, policy experts, journalists, politicians, administrative practitioners, and involved citizens who engage in an ongoing discourse about policy matters in a particular substantive area, i.e., health, poverty, environment). The goal would be to study the ways in which its members share background assumptions about the specific problem areas, their ideas about the relations of a particular science to decision making, the role of citizen involvement, and how they respond to outside opposition (Yanow 1996, 1998).

A field of argument can also be approached by studying how it is organized or distributed across a particular policy institution and its practices (Bourdieu 1977a). Policy organizations, those entities that policy analysts most typically serve, can be understood as arenas of policy argumentation.[3] As structures designed to fit intentions to practices, such organizations are animated by practices harnessed to mundane realities. Their rationalities lie in the concrete cases in which knowledge is created, used, and changed. They differ because they are functionally fitted to different aims, methods, and contexts. As Willard argues, however, in one sense or another, each field has its own sociology of knowledge. For this reason, we need to study these organizational bases of knowledge. This would involve an ethnographic examination of how such organizational actors go about their discursive interactions.

Basic to policy epistemics are the interrelationships between the empirical and the normative, the quantitative and qualitative inquiry. Whereas traditional policy analysis has focused on advancing and assessing technical solutions, policy epistemics would investigate the way interpretive judgments work to produce and distribute knowledge. In particular, they would focus on the movements and uses of information, the social assumptions embedded in research designs, the specific relationships of different types of information to decision making, the different ways arguments move across different disciplines and discourses, the translation of knowledges from one community to another, and the interrelationships between discourses and institutions. Most important, policy epistemics would involve innovating methods needed for coordinating multiple discourses in and across institutions.

In such a view, the very possibilities of formal analysis in decision making can be understood in terms of negotiations in normative spaces defined through controversies (Joss 1998). Controversies, in this respect, are not possible complements to empirical analysis but "governing processes" of formal assessment. In the words of Cambrosio and Limoges (1991), "Controversies define the degrees of freedom, and the conditions for the effectiveness of analysis."

Fundamental to such a perspective is a more thorough understanding of how socially contentious issues become the focus of public attention, how they are perceived and dealt with by the different participant actor groups, and how they are eventually settled through forms of bargaining and negotiation. In this understanding of controversies, the importance of various local, institutional, and cultural settings becomes apparent, as does the often disregarded but crucial difference between how science and politics resolve matters of uncertainty and controversy. Uncovering the epistemic dynamics of public controversies would allow for a more enlightened understanding of what is at stake in a particular dispute, making possible a sophisticated evaluation of the various viewpoints and merits of different policy options. In doing so, the differing, often tacitly held contextual perspectives and values could be juxtaposed; the viewpoints and demands of experts, special interest groups, and the wider public could be directly compared; and the dynamics among the participants could be scrutinized. This would by no means sideline or even exclude scientific assessment; it would only situate it within the framework of a more comprehensive evaluation.

In the case of environmental policy and risk analysis, as we have seen, these are just the kinds of concerns and problems that decision makers confront. Although grappling with the reliability of knowledge claims and the credibility of advocates is common to all fields, the problem is especially chronic in the field of environmental policy. That failures in environmental policy making can often be attributed to simplistic technocratic understandings of these relationships is clearly seen in the case of hazardous technologies and the politics of NIMBY. It is to these kinds of cultural rationalities underlying citizens' understandings and responses to expert advice that policy epistemics would turn our attention. They would help to make clear not only why citizens are hesitant to accept the authority of the experts but also how that knowledge gets translated and processed in the citizens' interpretive community. Such knowledge would help us better understand the ways in which the various players react to the scientific uncertainties that plague such policy areas. Epistemics also offers the possibility of finding ways around the political standoffs characteristic of NIMBY. Whereas policy epistemics might not offer us policy solutions, at minimum they could show us the ways to "keep the conversation going" (Rorty 1979).

In this view, an epistemic approach would connect risk assessment more directly to the public controversies that it seeks to inform. In terms of its uses, methodologies, and timing, such an approach would make available to decision makers a more in-depth and transparent characterization of the nature of public controversies than either conventional assessment methods or media debate can offer. For the purpose of policy deliberation, risk assessment would not only constructively open up the full complexity of such public controversies but also feed its results back into the evolving controversy. In the process, it would contribute to informed public deliberation, including the eventual closure of debate.

Central to such analysis would be the ways in which risk assessment is influenced by the processes of public controversy. Indeed, risk assessment was introduced by a polity in significant part as a political response to environmental and technological controversies. Important also is the recognition that the mode of deliberation, decision making, and conflict resolution in politics is not as different from that in scientific inquiry as scientists have led us to believe. In fact, science and its expert communities are themselves forms of political communities. In light of this awareness, there is little reason to presume politics in science is (or should be) handled much differently than it is in other areas of inquiry.

Policy studies has almost totally neglected this epistemic translation involved in policy making. And this neglect has occurred at considerable cost. Many of the most important failures that the discipline has confronted are directly attributable to this neglect. Most significant for present purposes is the issue of the relation of citizens to experts. Policy epistemics, for this reason, should be seen as a major challenge for the development of a more relevant mode of professional practice.

Conclusion

I began this work by citing John Dewey's challenge to the practice of democractic governance in a complex industrial society: How can lay citizens participate in making decisions about the increasingly complex social and technical problems of an advanced technologically driven society? How can citizens deal with issues so obviously dependent on the knowledge of experts? The answer for Dewey was a collaborative division of labor between citizens and experts. Toward this end, he called for an improvement of the methods and conditions of debate, discussion, and persuasion. Public debate would require the participation of experts, but rather than merely analyze and render judgments per se, they would interpret complex issues in ways that facilitate citizen learning and empowerment.

Modern expertise, as we have seen, has taken a different course. Throughout the twentieth century, scientific and technical experts have in effect largely set themselves off from the general citizenry. The result, as spelled out in the first four chapters, has been an increasingly technocratic form of public decision making. But the problem has scarcely gone away. To the contrary, it has only become more chronic. From the foregoing analysis, however, there is no reason to believe that these technocratic outcomes are necessarily inevitable, as many would have us believe. Indeed, the emerging practices of participatory inquiry that I have examined appear to be the methodological extension of Dewey's call for a more collaborative relationship between citizens and experts. In this respect, the contemporary problem seems to be less a question of methods than one of the political will to introduce and experiment with such practices on a larger societal scale.

For a democratic government, the outcome of this struggle with expertise is crucial. Citizen participation is the raison d'être of democracy. Not only does it give meaning to the term, but it plays an important role in legitimating both policy formulation and implementation.

In this work, I have sought to extend the role of citizen participation to policy-oriented science as well. Drawing on participatory experiences in environmental policy, I have shown that many citizens are much more capable of grappling with complex technical and normative issues than the conventional wisdom would have us believe.

But none of this has been to suggest that there is anything simple or obvious about such participation. In concluding, it is important to stress that throughout the discussion of environmental participation, I have sought to be clear about the problems associated with citizen involvement in the deliberative process. Although participation is a political virtue in and of itself, it offers no magic solutions or cure-alls. It plays an important role in effective policy making, but it is not the answer alone. Moreover, in practice, participation is a challenging and often frustrating endeavor. Beyond a good deal of democratic rhetoric, collective citizen participation is not something that just happens. It has to be organized, facilitated, and even nurtured. Without such commitment and concern, such efforts run a high risk of failure.

This concern brings us to the central political question that has informed a good deal of this discussion: Given that we cannot simplify the environmental issue, how can we innovate new relationships between citizens and experts that facilitate a wider range of lay participation? The question does not imply that all citizens need to participate all the time but does suggest that it would be much healthier for Western democracies if greater numbers were involved (Evans and Boyte 1992). In terms of democratic government, such participation is essential for developing citizenship, the cornerstone of democracy. For policy making more specifically, it is necessary for establishing a better connection between empirical investigation and citizens' local knowledges.

Toward this end, we have seen that experts have already innovated numerous techniques for facilitating the citizen's role in inquiry processes. The more critical question is how to bring these practices together in the professions. One of the important points in the discussion of participatory inquiry has been the recognition that this involves more a question of attitudes and practices than matters of scientific methodology. Here we confront a matter of power relations as much as an issue of knowledge production.

It is also important to emphasize that I have not argued for the use of these techniques in all cases. Rather, I have suggested that they should be available in the analytical toolbox for cases that benefit from increased citizen participation, a question itself for empirical inquiry. Participatory inquiry offers one such model for local problem solving. The consensus conference model provides a method for lifting such investigation to the broader societal level of deliberation. Not only do such practices vitalize environmental democracy in an age of expertise, but they can facilitate the kind of consensus essential to an effective strategy for dealing with our pressing environmental problems. Given the conflictual nature of these problems, such citizen inquiry should be explored as a political mechanism for developing and legitimizing a sustainable course of environmental remediation.

For the professional disciplines, this poses the challenge of rethinking the professional-client relationship. In such cases, the professional must learn to adopt more cooperative and facilitative interactions with the citizen-client. This, as we have seen, shifts the professional role from that of authoritative adviser to facilitator of client discourse. For this practice, professionals must develop a quite different set of skills. Rather than just to offer packaged solutions, the facilitator's task is to conceptualize and present policy alternatives and arguments for public deliberation. Beyond a competent grasp of empirical-analytical skills, he or she requires as well the ability to effectively share and convey information to the larger public. In this sense, the analyst is as much an educator as a substantive policy expert. The pedagogical task is to help people see and tease out the assumptions and conflicts underlying particular policy positions, as well as the consequences of resolving them in one way or another. Given the diversity of contexts and situations that have to be addressed in this assignment, the job has to be grounded in the interpretive skills of policy epistemics. In this sense, the analyst has to become skilled in task mediating across interpretive communities.

None of this, of course, will come about easily. The neopositivist methods of the social sciences remain deeply embedded in the standard practices of professional conduct. The issue of citizen participation in inquiry is perceived by many professionals as a threat to their status and authority. But the task poses an interesting and important challenge for the growing numbers of professionals who perceive the limits of the traditional methods. Participatory democrats within the professions should place the working through of these epistemics and their institutional implications at the top of the research agenda. Whether we are talking about large or small numbers of citizens (e.g., a political party or an advisory group), the prospects of a vigorous democracy in a complex society may well depend on it.

Appendixes

Appendix A: Risk-Benefit Analysis

The methodology of risk-benefit analysis is fundamentally an integration of two methodologies: risk assessment and cost-benefit analysis. The first method, risk assessment, is employed to evaluate risk resulting from both hazardous technologies and toxic health threats. Although the principles are the same, the assessment procedures are applied somewhat differently, depending on whether the focus is on technology or health. Because the discussions about risk-benefit analysis in this book primarily refer to toxic emissions from hazardous wastes, the methodology presented here applies to toxic exposure (Covello 1993).

The goal of risk assessment is to accurately predict the health implications of a hazard before or after it exists and to establish valid safety standards to protect the exposed population. The methodology typically specifies four interrelated steps (1) a process of hazard identification (e.g., Does a waste incinerator emit dioxins or heavy metals?); (2) an assessment of human exposure (e.g., Can the various routes of the toxin to the affected population be traced and how much of it enters the human body?; (3) the modeling of the dose responses (e.g., What is the empirical relationship of the exposures to the chemical under investigation and the frequency of adverse impacts?); and (4) a characterization of the overall risk (e.g., How does the data as a whole provide an overall evaluation of the toxic implications for human health, most commonly defined in terms of cancer?). In an effort to err on the conservative side of safety, risk assessors most often use "worst case scenarios." The overall risk is generally expressed as the probable number of cancers per million people who are exposed over the course of a standard life expectancy.

Basic to risk-benefit analysis are questions about the ability of risk researchers to quantify accurately the particular risks, especially given a general scarcity of empirical data about chemical effects and the nature of the assumptions about exposure and responses that guide the assessment process. With regard to quantification, the assessment of exposure is especially complicated. Here the evaluator attempts to construct a sophisticated statistical model based on simulations of the movement of the hazardous substance (e.g., through air, water, and animals), and on estimates of human activity that would create exposure to it, along with hypotheses about how the substances actually get into the human body. In particular, disputes emerge over the extrapolation of findings from high-dose experimental settings to low-dose real-world circumstances, over the comparability of short- and long-term exposures. (Other disputes raise questions about the use of animals in testing.) Disputes also arise over the question of which health outcomes the researchers should concentrate on. Most commonly, risk assessors limit their focus to cancer and ignore other detrimental effects to the human immunological, reproductive, and nervous systems.

The second phase of the risk-benefit analysis is the cost-benefit analysis (Crouch and Wilson 1982). Here the goal is an explicit comparison of the benefits derived from a hazardous activity and the risks that are involved in that activity. The costs, however, are defined in terms of specific levels of risks rather than monetary value. The method thus involves calculating the benefits of a project (adjusted against regular costs, such as plant construction and maintenance costs), comparing the ratio of the risks to the benefits, and multiplying the resulting figure by the total number of people affected. For example, it might be discovered that a power generator located in a particular community would spew toxic chemicals into the air that would lead on average to one death for every million local residents per facility per year and would offer power for $0.11 per kilowatt hour of electricity. Another type of generator, it might be determined, could lead to an average of two deaths per million community members per facility, but would offer power for $0.08 per kilowatt hour of electricity. For the risk-benefit analysts, these two types of impacts — deaths per million and price per kilowatt hour — are said to be "objective categories," as their actual levels are taken to be empirical facts (Hiskes and Hiskes, 1986, 177).

Appendix B: Alternative Dispute Resolution

A variety of alternative dispute resolution methods have been developed for shaping consensus among administrative agencies and the relevant groups involved in environmental conflicts. The dominant variant of the practice involves a form of mediation designed to facilitate a consensual agreement among the parties to a dispute. Mediation, as the National Research Council (1996, 201) explains, "requires the involvement of the spectrum of interested and affected parties for any agreement to be implemented without determined opposition," an idea resting on the belief that "unless the parties feel they have affected the decision, it is not likely to be satisfactory to them."

As a rule, there has to be a willingness on the part of all parties to seek an agreement. If there is entrenched opposition on the part of a particular party, the technique is usually of limited value. In general, the goal is to reach a formal settlement of the dispute or conflict.

Practiced at all three levels of government, the negotiation process has taken place in the various phases of the policy or regulatory process, from the formulation of laws to their implementation. After the parties have worked together to propose legislation, the proposal is submitted to the normal processes of public comment and review. The practice has often been used by the U.S. Environmental Protection agency to formulate complex technical rules, especially in instances of high degrees of uncertainty. The objective is to reduce the number of legal challenges that have typically followed EPA's rulings by involving the adversaries in the decision process.

The parties selected to participate usually represent the important interest group participants. Because they frequently must decide highly technical issues involving gaps in theoretical and empirical knowledge, groups are sometimes chosen for their relevant expertise. A group's success can significantly depend on its abilities to engage in persuasive argumentation.

The National Research Council (NRC) (1996, 202) compares the strengths of alternative dispute resolution (its ability to "deal with complex issues, strongly held beliefs, polarized opinion, conflicting values, and technical concerns") with its weaknesses (the parties' willingness to accept the practice over direct action or litigation). Other questions that are addressed by the NRC include whether the right participants have been brought to the process, is the power balance among the groups fair enough, and whether there is sufficient commitment to the negotiation process?

The overall assessment of alternative dispute resolution is mixed. Some policymakers find more flexibility and trust has developed among the groups as a result of engaging in these processes. But in other cases, the result has been an increase in the level of conflict. Such failures are often attributed to critical issues or needs that have been obscured in the negotiation process, often attributed to power imbalances among the participants (Crowfoot and Wollendeck 1990; Baughman, 1995).

Appendix C: Farmers as Analysts, Facilitators, and Decision-Makers in Participatory Resource Development Programs

In most development programmes farmers are informants or at best data collectors. They do not participate in analysing and taking decisions based on the analysis nor is their inherent analytical capacity used. Although the resultant development process may lead to tangible development results in the short run, it does not encourage sustained innovation by the local villagers or institution building at the village level.

In contrast, the Aga Khan Rural Support Programme (AKRSP) in its work with village communities [in India], has tried to involve villagers in collection, analysis, and use of data, and as facilitators of a participatory appraisal and planning process.

The Aga Khan Rural Support Programme is a non-governmental organization established in 1985 to promote and create an enabling environment for the village communities to manage their local natural resources in a productive, equitable, and sustainable manner through their own village institutions. The villagers are encouraged to develop a local cadre of village extension volunteers who develop expertise in appraisal, planning, implementation, management, and monitoring and also build functional linkages with other state, nongovernmental, cooperative, and financial organizations in the area . . .

The emphasis is not on creating a large support organization which has expertise in all functional areas, but on encouraging villagers to volunteer to become village para-professionals in different areas depending on their interest and aptitude. Participatory Rural Appraisal (PRA) is used as a major training and planning methodology to enable village volunteers to become village analysts, managers, and institutional change agents . . .

In the process of using Participatory Rural Appraisal, participatory mapping by the village community has emerged as a key method to enable village communities to engage in problem-solving, analysis, appraisal, planning, and decision-making. Maps are prepared on the ground using a number of local materials such as stones, seeds, twigs, and local colours. Use of these symbols enable a number of illiterate and inarticulate people in the community such as women and landless people to participate . . .

[The] maps indicate the majority of natural resources in the village including land and water, local land-use classifications, and the catchment and command area of each resource. It also shows the quality or status of each resource and its likely users. People also indicate[d] qualitative data regarding use of these resources. The following types of resources maps are . . . prepared by the villagers. (i) An inventory of the village's natural resources in terms of local land-use classification[s] . . . ; (ii) A map of the existing status of resources . . . ; (iii) Maps showing the utilization of various resources in the village . . . ; (iv) Maps showing the uplands, midlands, and lowlands in the villages and the characteristics of these land type[s]; (v) Maps showing the quantum and extent of the resources . . . ; Maps showing the users of various resources, e.g., a community well, common forest land . . . small stream[s], lowlands, and drinking water village pond. . . .

Maps focus discussion and lead to a sound basis for trying out other analytical methods like transect walks and focus group discussions. This ensures that more people participate effectively in the discussions and have a common framework for further discussions. The maps also provide a check-list to ensure that issues identified at the start of the project are not missed out in later stages. . . .

In heterogeneous societies with a number of caste, social, and economic groups, it is important to know the stratification of the communities both in terms of resources and their access and distribution. These have implications for the solutions and their likely applicability. These maps show the distribution of households in terms of different caste and social groups in the village. The social map is then extrapolated with other resource and thematic maps and the problem identified can be correlated with the social aspects. This is linked with ownership of assets and wealth groupings for the village. Making such a map helps to analyse how each solution identified by the community affects different social groups and particularly the poor. After more experience with this kind of analysis people can cross-reference social maps with other maps to understand the social implications of their existing endowments and the solutions identified by the community. These maps are an important mechanism by which social and equity analysis becomes an integral part of the appraisal process by the community. . . .

Farmers' analytical skills have . . . [been] enhanced in a number of ways. [They] make line diagrams about technologies they were trying out in different zones of the village land. . . . These were presented to other farmers resulting in an inventory of local technical innovations or, where inadequacies were identified, the incorporation of further suggestions from outsiders. A plan to test out the impact of these innovations on problems associated with particular land or soil types was set out in a line diagram which was used to discuss the ideas with men and women farmers in the field. . . . [In addition] farmers . . . carr[ied] out monitoring and impact studies using a range of participatory methods. . . .

Further analytical skills developed by farmers concern cost-benefit analysis and equity mapping. . . . [The] information is shared with all the village community members. The form of presentation enables villagers to understand viability aspects which are difficult . . .

[Another] skill acquired by farmers [is] as facilitators for the participatory planning process. . . . [Moreover] farmers [make] a presentation of a plan for the long-term management of their natural resources to a team of [district officials]. . . . The officials and villagers . . . split into . . . teams and the villagers show . . . them a number of problems and solutions proposed by them in the plan (Parmesh Shah 1995).

Appendix D: Community-Based Participatory Research
The North Bonneville Powerplant

[The] U.S. Army Corp of Engineers, after analyzing eleven potential sitings for a second energy powerhouse at the Bonneville Dam on the Columbia River, announced that the best location for the powerhouse (and a new channel in the river) would be the center of the town of North Bonneville in the State of Washington. Every family in the town, and many around it, faced eviction and relocation. With a population of 470, including one-third on fixed income and about 40 percent unemployed or only seasonally employed . . . the town seemed destined to be another insignificant footnote in the history of towns destroyed by Corps of Engineers projects. . . .

When the Corps of Engineers told the town's residents . . . that they were going to have to move . . . the town found itself rallying around a common desire and goal — relocation as a community, where social bonds . . . would be maintained. They formed North Bonneville Life Effort (NOBLE) and, with the assistance of the Bureau of Community Development of the University of Washington, completed a survey documenting that 64 percent of the residents preferred to relocate into a new North Bonneville as close as possible to their existing town. . . .

In its search for assistance in maintaining its identity, the town contacted the Evergreen State College in Olympia, Washington. . . . The principal faculty member, Russell Fox, had been strongly influenced by his participation in a Chilean government program designed to decentralize Ministry of Housing and Urbanization decision making, and by his role as organizer of a participatory research and land use planning project [in] a semi-rural community in Washington state. Fox and Evergreen students were looking for projects that would demystify the planning process, decentralize community structures and decision making, and empower citizens through participatory research. The residents of North Bonneville discovered the faculty and students and a four-year participatory research project was underway. . . .

The students quickly discovered that, although town's residents had extensive knowledge about their community and strong feelings and desires concerning their pending relocation, they did not understand the complex external political and social forces that were determining their future. . . . After discussing the general nature of the town's problems, goals, and commitment to active participation in the planning process, Fox and the students made [a] proposal for a participatory research process. . . . The general strategy [was] for the students to live and work in the community, while creating and gathering the quantifiable planning data needed and engaging in ongoing discussions with residents so that the residents could create and discover their own understanding, expression, and use of the data. The data with which to make the decisions, an awareness of the external forces affecting decisions in their lives, and the self-confidence and capacity to make their own decisions all [were] developed simultaneously. [Guided by Fox], the students through informal discussions with residents, community workshops, and internships with principal agencies, would (a) share with the residents what they were learning about communities, the planning process, and the skills of participation; (b) compile, organize, and publish a report with pertinent information available about the relocation problem and the characteristics of the existing community; and (c) take leadership . . . interested in the relocation issue, while involving the town in discussions that would lead to political skills and strategies they could use in pursuing their goal.

The town, initially through actions of its elected council, would (a) provide work space and assume some of the costs of travel and living expenses for the students, (b) actively promote the participation of the residents, and (c) assure that the entire process be genuinely open to all groups and members of the community. The . . . outcomes of the six-month project would be (a) the publication of a relocation planning study that the residents would understand and be ready to use; (b) an increased awareness on the part of all residents of the nature of their community, the nature of the relocation problem, and the options available to them, and (c) an increased readiness to politically participate in the pursuit of their goal to relocate into a new North Bonneville. . . .

[As the project] began the students welcomed residents dropping in [to their storefront office] to help add figures, to locate features of their community on maps, or to describe in detail no outside researcher would ever identify, the social networks among community members or the special ponds or groves that different residents had claimed as their special places to fish or picnic. The students met with residents in their homes, in the post office, in the cafe, and on the river bank. The discussions were dialogues of sharing what they each knew of the community and its problem. The residents knew their community, how it worked, and what was special about it. The students shared what they were learning about the planning process; about the relocation laws; and about the technical data they were developing, or discovering through research (often involving a resident or two) into information available through different state or local agencies. In a few cases the students . . . spent time working as interns in [various relevant state and local agencies]. More formalized contacts between the students and residents included scheduled coffee hours in homes where the particular relocation situation and options of each family could be discussed, weekly workshops and presentations of information being generated which were open to the whole community, and almost daily contact with the staff and elected official of the town.

As residents reflected and talked about what they knew about their community, they began to realize the [discrepancies] between their knowledge of who they were and the very different perspective of the Corps of Engineers and the politicians who wrote relocation laws. They began to define their community as a complex network of social, natural, and spiritual relationships. They discovered that the government defined their community as abstract individuals and a quantifiable number of physical artifacts, such as a fire truck and so many lampposts. Similarly, as the residents learned about planning processes — both those imposed by external forces and those they were creating — they realized that planning was merely the creation of information to implement goals. They realized that the Corps's planning was a meticulously designed and carefully controlled critical path for technical efficiency. To the contrary, the town's "process planning" was the creation of knowledge about themselves, including the potential for implementing their own rather than the Corps's goals. They discovered that their goal — survival of the social relationships that defined their community — was quite different from the government's goal — to build a [power-generating] plant as quickly as possible.

These discoveries and the students' persistent encouragement and affirmation of the town's ability to control the situation gave the town's leaders the confidence and skill to act on their own perception of reality rather than be limited to fighting the Corps's perception of reality through the channels the Corps controlled. [As the community discovered], the members of the Corps of Engineering, [as] masters . . . of construction logistics and pursuit of . . . planning goals . . . can out-professionalize anyone who challenges them at their own game. The Corps of Engineers were continually baffled and outmaneuvered, however, when faced with an entire town of residents who weren't represented by professionals but who knew the data, information, and processes themselves.

[One of the] major examples of the Corps inability to understand or control this development of popular knowledge was the Corps . . . reversal of their earlier claims about not being able to fund planning for a new town. Once the town's planning effort began gaining momentum, the Corps came to the town with a slick presentation of a planning process they would undertake for the town. The town's residents listened politely and responded with a firm "No, thank you." They recognized that they and the students were already doing everything that the Corps proposal included and that the town, rather than the Corps, was a client and owner of the study. [In the processes], the town council came to realize that one of the Corps's most effective strategies was to control and manipulate information and keep united fronts from forming by selectively telling different agencies and segments of the community different information. . . .

As one of their programs . . . the student group began working with different segments of the community on the initial conceptual characteristics and relationships that would lead to the design and layout of a new town . . . that reflected their life-styles, values, and social and economic relationships. The design of the town reflects, far better than the original town did, the residents' relationships to each other and their physical environment. [Moreover], the town has taken the increased consciousness of itself and its potential into new areas of learning and action. For example, the town secured grants for pilot drillings to explore the potential of geothermal energy as a source of the town's heat. The pilot wells were successful, [leading the town] to pursue public and private capital to install the country's first community-wide geothermal heating system . . .

[Comstock and Fox (1993) see the experience offering a number of lessons]. . . . It demonstrates the potential for participatory research to provide a basis for successful political struggle by a community. [They argue that it provides a model for] other communities . . . faced with economic and social destruction at the hands of government agencies or private organizations . . . [It] also shows that participatory research can initiate a sustained process of political organization by a community along with the personal growth of its residents. The people of North Bonneville . . . [continually learned] about themselves and their environment. . . . They put this knowledge to use in creating a new community that preserves such progressive values of the old as a respect for the land. [The experience] provides an historical justification for the progressive social science represented by participatory research. . . . [It illustrates, they argue, that participatory research can provide the basis for a critical praxis long needed to accompany the development of a critical theory of society] (Comstock and Fox 1993).

Appendix E: Participatory Political Analysis in a Social Service Agency

Dan Durning (1993) presents a case study of a "stakeholder" approach to participatory policy analysis. What follows are excerpts from his analysis of how the Georgia Division of Rehabilitation Services used an 11-member team of agency employees to analyze the agency's "order of selection" policy and to present findings to its executive committee.

> In April 1989, the executive committee (EC) of the Georgia Division of Rehabilitation Services (DRS), a division of the State Department of Human Resources, commissioned an analysis of its order of selection (OS) policy, which sets priorities among potential service recipients. It asked its Planning and Research and Special Projects (PRSP) sections to lead a policy analysis project team consisting of employees throughout DRS.
>
> The appointment of an employee project team, instead of a single analyst, to conduct the policy analysis was consistent with DRS's style of operation. In the 1980s, it had established itself as a "participative" organization. Its major policy decisions had been turned over to its executive committee, which was made up of the heads of the agency's divisions. And, in its long-range plan, updated annually and given serious attention, "the importance of consumer and employee involvement" is one of the six elements of the "philosophy of organization." To involve consumers in its decisions, DRS developed a strategy for obtaining advice from the people who use its services. To involve employees, DRS institutionalized a system of employee "study groups" and task forces to examine key issues facing it.
>
> DRS provides financial and other assistance to disabled people who need such aid to obtain jobs. Before doing so, DRS counselors must determine if applicants are eligible for assistance and if they fall into an OS disability category for which funds are available. If an applicant is due assistance, a DRS counselor helps him or her prepare a plan to prepare for employment. Then, DRS pays for the assistance specified in the plan. . . .
>
> In April 1989, the DRS executive committee approved a proposal by the PRSP for a policy analysis of the OS. Shortly after the approval, a project team was assembled to the study. It included: A PRSP staff person as the team leader–policy analysts (TL–PA); two assistant district directors (ADDs);

three counselors (added after the first meeting); two unit supervisors; three ad hoc members, including the chief of the policy unit, a quality assurance specialist . . . and a senior operations analyst, to obtain information to be used in the analysis. . . .

At [the] first meeting . . . the project team discussed the general framework for a policy analysis as described in relevant textbooks. To guide its work, it decided to use a "textbook" framework that began by defining the problem and ended with making a choice. . . .

At this first meeting, the team members discussed the type of information needed to conduct the analysis, and they divided up the task of collecting it. The TL–PA volunteered to collect information about the OS policies of other states. The unit supervisors and ADDs agreed they would, during the course of their normal work, assess the degree of compliance with the current OS policy. The central-office team members were asked to provide a large amount of statistical information relating to the OS policy.

At [the] second meeting . . . the project team members discussed possible problems with the OS policy based on their experience with the policy. Also they proposed some criteria to use when comparing alternatives . . . and they reviewed the information about the OS policies of other states and relevant DRS data extracted from the agency's data sources. Finally, they winnowed a long list of issues to be studied to five key questions they would investigate.

The team members were asked to take those key questions back to their workplaces to see how they were answered there. Also, the TL–PA volunteered to raise the questions at a meeting of the ADDs to get their views on them. In addition, the central staff members on the team were asked to provide additional quantitative information on the composition of DR clients by OS categories.

In their July meeting, the team members decided their informal survey did not provide adequate information for their analysis. They decided to survey district directors, ADDs, and counselors to obtain other views on problems associated with the OS policy. This written survey was conducted by the TL–PA, and the results were compiled before the next meeting.

Also at the July meeting, the team members anonymously submitted their suggestions for changing the order of selection policy. Those suggestions were compiled, distributed to members, and discussed. They formed the basis of the alternatives considered in subsequent meetings.

By the end of the September session, after over 16 hours of meetings, the team members had many pieces of the first-cut analysis. They had identified several problems with the OS policy through their discussions and the survey. They had generated some alternatives. They had refined the criteria to be used to compare alternatives. . . . Also, they had collected and reviewed information and data about the OS policy; though they did not get all the data they wanted, they had a substantial amount to help them with their problem definition and comparison of alternatives.

The next two meetings . . . were spent refining the previous work and predicting the consequences — positive and negative — of each of the alternatives they were examining. The remainder of the team's meetings were devoted to polishing its earlier work and putting the analysis on paper. The different parts of the analysis were contained in several draft reports by . . . the TL–PA, based on the discussions and decisions at the team meetings. [The TL–PA] circulated drafts of the report to team members and, during the later meetings they discussed, almost line by line, the draft report. The final report reflected the consensus policy analysis of the group.

The project team offered the EC three alternatives to the existing OS policy and predicted the consequences of each. Many team members believed that the team should not make a choice from among its alternatives because the choice could not be based on such criteria as efficiency or effectiveness, which they could judge. Instead, the choice would reflect their values, and the team members felt that the EC should make such value judgments.

The project team's report was submitted to the EC in April 1990. The EC members received it positively and, in interviews, they expressed satisfaction with the quality of the report. . . .

Compared to traditional policy analysis, this organization-stakeholder policy analysis appears to have the following strengths:
The team thoroughly understood the context of the analysis. . . .
Team members were a good source of opinions, data, and information. . . .
The team had the resources to construct a "mental model" to predict the consequences of the proposed alternatives. . . .
The process of analysis created spin-off benefits. . . .
The analysis had credibility within DRS. . . .
The project team approach had several weaknesses and potential problems that decreased its value as a model of policy analysis to be used more widely.
The analysis was slow and removed employees from their regular jobs. . . .
The cost of the analysis was substantial. . . .
The analysis may have used less sophisticated methods than a technical policy analyst would have used. . . .
There are dangers of organization breakdowns. . . .
Some issues may have been ignored to get consensus advice. . . .
Failure to use policy advice may discourage future participation in stakeholder policy analyses.

Based on this case study, I would conclude that organization-stakeholder policy analysis is well suited for addressing some messy or ill-structured policy issues; these types of issues are defined by Dunn as "decision problems . . . for which decision makers, preferences or utilities, alternative, or outcomes, or states of nature are unknown or equivocal." For the analysis of such policy issues, the technical methods of traditional policy analysis are inadequate, and "second-order" methods of analysis are needed. State holder policy analysis qualifies as a "method of the second type."

In many cases . . . a policy issue is part of a complicated context in which ends may not be well defined, the definition of problems may not be settled, the meaning of data may be disputed, and the legitimacy of proposed policy instruments may be the subject of internal debates. [S]takeholder policy analysis may be valuable because the stakeholder team can negotiate an understanding of the context of the decision and can transform inputs into advice using a model to predict the outcome [of] different alternatives. [Durning 1993]

Appendix F: Deliberative Experiments

During the past decade, there has been an impressive elaboration of techniques designed to find what citizens think about policy issues. In large part, these efforts have been developed to deal with the fact that the most widely used technique, the public opinion survey poll, introduces numerous biases that obscure what citizens in fact think (Lindeman 1997). Most problematic, standard polling techniques offer citizens little or no opportunity to reflect on the questions put to them, especially with regard to those questions that involve unfamiliar and uncomfortable ideas. Such polling techniques cede agenda control to the survey designers, who assume they not only know the right questions and answers but how to interpret citizen responses. In addition to structuring the agenda for the discussion, survey designers determine which facts and arguments the participants should respond to.

Deliberative research focuses in particular on efforts to extend the role of citizens in setting the agenda for such inquiries. Concerned with what Fishkin describes as "considered judgment," such work seeks to understand the processes citizens engage in to arrive at informed, responsible public preferences (Fishkin 1996). Toward this end, research has followed two particular lines of investigation, poll-oriented and group-oriented deliberation. On one end of the spectrum are those experiments that have tried to improve on polling techniques themselves. For example, the American Talk Issues

Foundation (1994) uses telephone interviews to mitigate the problems of polls by giving respondents more time and resources to assist them in thinking about their answers. This method is based on the idea that if policy choices are clearly worded, basic textual information provided, and key arguments on both sides clearly presented, people can often come to considered judgments in a few minutes. Such deliberative telephone polls attempt to mirror more closely public opinion through large samples (1,000–1,500) respondents. The approach is clearly an advance, but still remains subject to the main criticism against polls — the designers play too great a role in the process.

A second innovation is the "televote." In this approach, participants (solicited by random-digit dialing) receive an informational brochure with basic background information, varied expert opinions, and policy alternatives. Not only are they encouraged to take as much time as they need to read the materials, they also discuss them with other people before casting their "votes." Some televotes have been conducted in conjunction with "Electronic Town Meetings" that allow larger numbers to participate. The main criticism of the televote is that, like any forced-choice survey technique, it leaves to the designers control over the agenda of questions and language in which the responses must be given.

A third technique is the focus group. Led by trained moderators, participants meet in small groups of about 25 to discuss the issues for about an hour. The process usually begins with supplying the participants with information (provided in both written and audiovisual formats) designed to inform debate on a small "menu" of broad policy choices. The ensuing deliberation is structured to ensure that participants have considered the pros and cons of each choice. At the end of the process, participants state their opinions, not only on the broad choices, but also on various narrower policy issues. The Public Agenda Foundation, for example, typically gathers large (representative) groups — preferably several groups in various cities — for moderated focus-group discussions on a policy issue (Immerwahr and Johnson 1994). In one such project, 800 people across five cities were assembled. Criticisms of this method emphasize the limited amount of time participants have available for deliberation. An hour or two of discussion is arguably long enough to grasp the outlines of how core values might apply to the crucial choices at hand, although certainly not long enough to do much sustained thinking. Moreover, the discussions are also structured in a way that provides participants with little control over the agenda.

One of the most discussed approaches is that of Fishkin's National Issues Convention. Fishkin's method seeks to combine both deliberative polling and small group deliberations (1996). For example, the National Issues

Convention gathered 450 citizens from around the country to meet for four days of discussion on crucial national issues, both domestic and foreign. Participants assembled in small groups to talk about their positions, listened to experts and presidential candidates, and had limited opportunities to ask questions of those witnesses. Designed to combine the depth of small groups with the rigor of sophisticated polling techniques, the breadth of the agenda militated against depth.

While the examples above all illustrate the effort to get beyond the one-dimensional survey techniques that have largely dominated the collection of citizen preferences, none achieve the kind of citizen involvement permitted by citizens' juries or panels. Citizens' panels are first and foremost an effort to provide citizens with an opportunity to deliberate in some detail among themselves before coming to judgment or decision on questions of public policy (Crosby, Kelly, and Schaeffer 1986).

The concept of citizens' panel first emerged in northern Europe, although the practice has now spread to a range of countries around the world, including the United States, where it is generally known as a "citizens jury" (National Research Council 1996, 203–4). One of the most elaborate formulations of the method has been developed by Peter Dienel of the University of Wuppertal. Dienel's concepts of *Buergergutachten* (citizens' assessment) and *Plannungszelle* (planning cell) are put forward as a way of addressing a "deficit of legitimation" resulting from the isolation of experts from citizens in "establishment democracies" (Dienel 1992, 10; 1989).

A typical citizens' panel or jury assembles twelve to twenty-four randomly selected citizens for three to five days to discuss among themselves a particular question or issue (Crosby 1995; Kathlene and Martin 1991). In the case of the Plannungszelle, all citizens in the relevant community who are eighteen years old or older have an equal chance of being selected. Moreover, during the actual period of the inquiry, the participants are exempted from their regular work obligations. They are either given a leave of absence for "continuing education" or compensation for financial losses they might incur as a result of their participation. People responsible for the care of others, in particular parents with small children, are supplied with day-care assistance (*Buergergutachten Uestra* 1996).

The Plannungszelle, which has spread beyond Germany to some twenty countries internationally, is often organized around a number of groups meeting at the same time. This means that up to 200 or more citizens might participate in the assessment of a particular topic. In the process, the panels or jurors are typically asked to express a preference for one of three or four policy options. A panel is assisted by a moderator who keeps the deliberation moving, although jurors have considerable discretion to determine their own agenda and procedural rules pertaining to schedule and moderation (e.g., time limits on individual and group exchanges). The moderator arranges for experts to be available during deliberations to help clarify questions of fact, with jurors being encouraged to actively interrogate the answers of the experts. Because jurors have considerable time to hear expert testimony and to ask questions, they can learn much more about an issue than is the case in other deliberative forms (e.g., deliberative telephone polling or televoting).

At the end of their deliberations, a report on the panel's findings and conclusions is prepared. Conflicts among jurors are generally resolved through the principle of majority vote, although minority views can be reported. In addition to the foundations or agencies that might have commissioned the citizens' panel, the report is sent to a range of bodies agreed to by the participants at the outset (*Buergergutachten Uestra* 1996).

The main criticisms of citizens' panels are that (1) the topics are usually framed by the organizers with a particular question of interest to sponsors in mind; (2) these problems concern rather narrowly defined local problems (e.g., how to reorganize public transportation routes); (3) the discussions among the participants are closed to outsiders; and (4) the report is often written by or with the assistance of the group moderator of the project. All four of these problems are corrected in the Danish consensus model, discussed in the chapter.

Notes

1. Democratic Prospects in an Age of Expertise

1   Giddens explains expert systems in the following words: "By expert systems I refer to systems of technical accomplishment or professional expertise that organize large areas of the material and social environments in which we live today. Most laypersons consult 'professionals' — lawyers, architects, doctors, and so forth — only in a periodic or irregular fashion. But the systems in which the knowledge of experts is integrated influence many aspects of what we do in a *continuous* way. Simply by sitting in my house, I am involved in an expert system, or a series of such systems, in which I place my reliance. I have no particular fear in going upstairs in the dwelling, even though I know that in principle the structure might collapse. I know very little about the codes of knowledge used by the architect and the builder in the design and construction of the home, but I nonetheless have 'faith' in what they have done. My 'faith' is not so much in them, although I have to trust their competence, as in the authenticity of the expert knowledge which they apply — something which I cannot usually check exhaustively myself" (Giddens 1990, 27–28).

2   One can, to be sure, point to the ways that the Internet and email have connected peoples all over the world. Some have even seen this as the basis for a new kind of civil society. But here the arguments of those critical of this view remain persuasive. When it comes to politics, Internet users remain isolated individuals with none of the social bonds or face-to-face interactions that provide the basis for a political movement or group. Thus far, in any case, there is no compelling evidence to

suggest the emergence of new electronically based forms of decision making; that is, nothing like the new interactive grassroots democracy suggested by Toffler and Gingrich. Indeed, one can argue that excessive involvement with the Internet and computer games retards the development of the kinds of social skills needed for effective participation in politics.

3   Mainstream economics is an especially important example of modern-day positivism. Because of its rigorous scientific orientation (and, less openly stated, its relations to power), economics is widely acknowledged to be the queen of the modern social sciences and is seen as the model to emulate. Basic to this much admired scientific rigor is the principle of the fact-value dichotomy. Economists labor to separate out all forms of subjective judgments, which are considered unsusceptible to scientific analysis and are largely defined as unmeasurable. Indeed, today it is possible to read an economic textbook without encountering the word "capitalism," a term seen to be problematically associated with class struggle between owners and workers. If one still needs a straightforward statement of positivism, it can be found in the popular textbook of Harvard economist N. Gregory Mankiw, *Principles of Economics*. Mankiw puts it this way: economists "make a distinction now between positive or descriptive statements that are scientifically verifiable and normative statements that reflect values and judgments. The question is, can you do positive economics without normative economics. I think so" (Mankiw quoted by Uchitelle 1999, B7).

4   Given Foucault's theoretical difficulties with the normative questions of agency, his theoretical approach offers little assistance with the particular questions of interest here.

2. Professional Knowledge and Citizen Participation

1   An expert is defined here as a person who has a high level of competence in a body of knowledge and the methods that generate and test such knowledge. Professional expertise generally pertains to a specific field of discipline inquiry and its practices, such as medicine, law, architecture, engineering, physics, psychology, or social work. The proficiency of an expert is typically certified in terms of the standards and practices of a peer group, usually organized in the form of a professional association. More specifically, this generally involves demonstration of knowledge of theoretical frameworks and the relevant bodies of literature, established causal knowledge and general rules,

skill in the use of particular instruments, proficiency in the processes of data collection, and the norms of professional conduct. Based on such validation, the professional expert's credential is given official recognition by the larger society. In most cases, it involves the granting of a license to engage in professional practice.

2   Another study by the Council for Excellence in Government in 1999 found that fewer than 40 percent of the American public believes that their government is "of the people, by the people, and for the people." Two-thirds of the respondents reported that they do not feel connected to their government.

3   Contemporary democratic systems, as Pateman (1979) puts it, do not afford their citizens enough acts of participation to generate the kinds of expressed consent needed to ensure political legitimacy.

4   "Democratic elitism," a term coined by Bachrach and Baritz (1961), refers to a thin conception of democracy that involves the rotation of a small number of elite groups in and out of top-level decision-making circles.

5   The term "participatory inquiry" refers here generally to a range of collaborative approaches to research. These various approaches and their practices are taken up in chapter 9.

6   The term "specialized citizen" is borrowed from Paris and Reynolds (1983).

3. Environmental Crisis and the Technocratic Challenge

1   Beck clearly exaggerates here, an unfortunate tendency that runs throughout his work. The problem is, it is just as easy to argue that we live in the "safest of times" as the "riskiest of times," which is itself an important fact of environmental politics. Beck could write *The Risk Society* in Germany in the middle 1980s, particularly given the anxieties Chernobyl unleashed on Europe, but it is hard to imagine such a book appearing in the United States. The same is true of England, where the book was first greeted with incredulity.

2   Insurance companies, it is worth noting, have already signaled their inability in the future to pay the costs of damages from hurricanes and other forms of turbulent weather caused by global warming.

3   Basic to the procedure is a comparison of accepted existing risks with new risks proposed for acceptance. In the case of favorable comparisons, the legitimation of the former is supposed to be transferred to the latter.

4 Beck offers no clear-cut definition of "ecological democracy." Dryzek, however, offers some assistance. In a discussion sympathetic to Beck's concept of reflexive modernization, Dryzek defines ecological democracy as "any enhancement of democratic values in an ecological context that does not sacrifice ecological values, or any enhancement of ecological values that does not sacrifice democratic values." In ideal form, it "would involve a "felicitous combination of the two" (Dryzek 1996, 108–9). Basic to the achievement of these values, in Beck's conceptualization, is the political process of reflexive modernization.

4. The Return of the Particular

1 There is no standard definition of "postpositivism." The conception of postpositivism presented here follows, in this regard, no particular school of thought. Rather, it represents an assimilation of contributions from social constructivism, informal logic and practical reason, discourse analysis, feminist epistemology, and the postmodern theory of knowledge. Most fundamentally, "postpositivism" is grounded in the idea that reality exists but can never be understood or explained fully, given both the multiplicity of causes and effects and the problem of social meaning. Objectivity can serve as an ideal but requires a critical community of interpreters. Critical of empiricism, "postpositivism" emphasizes the social construction of theory and concepts, and qualitative approaches to the discovery of knowledge (Guba 1990). McCarthy (1978) has defined the task of developing a postpositivist methodology of social inquiry as figuring out how to combine the practice of political and social theory with the methodological rigor of modern science.

2 The term "neopositivism" is used here to refer to the modern-day embellishments of "positivism." In most general terms, positivism is an epistemology — a theory of knowledge — holding that reality exists and is driven by laws of cause and effect that can be discovered through empirical testing of hypotheses. Such inquiry can be empirically objective and value free, as the laws or generalizations exist independently of social and historical context. Today positivism as a concept serves as much to fuel a polemic as it does to identify a distinct epistemological theory or movement. "Neopositivism" is employed to refer to the modern variants of positivism. As such, the term pertains to a legacy of concepts and theories, techniques, attitudes, and convictions that have their origins in positivism.

3 Underlying the effort to separate facts and values is a fundamental positivist principle, "the fact-value dichotomy" (Proctor 1991). According to this principle, empirical research is to proceed independently of normative context or implications. Because only empirically based causal knowledge can qualify social science as a genuine "scientific" endeavor, social scientists are instructed to assume a "value-neutral" orientation and to limit their research investigations to empirical or "factual" phenomena. Even though adherence to this "fact-value dichotomy" varies in the conduct of actual research, especially at the methodological level, the separation still reigns in the social sciences. To be judged methodologically valid, research must at least officially pay its respects to the principle (Fischer 1980).

4 From quantum theory and its postulate of indeterminacy we have learned that aspects of the atomic level of reality are so influenced (or codetermined) by other dimensions of the same phenomena that such processes can no longer be described as determinate or predictable. Moreover, such research has led some physicists to argue that the explanation of the behavior of a particle depends in significant part on the vantage point from which it is observed (Galison 1997). Chaos theory has demonstrated that an infinitesimal change in any part of a system can trigger a transformation of the system at large (Kellert 1993). Such empirical phenomena are thus defined better as "participatory interminglings" than as perceptions of objective things standing apart from human subjectivity. In short, the traditional understanding of the physical world as a stable or fixed entity is no longer adequate. For neopositivism, this poses a fundamental problem: it loses its firm epistemological anchor.

5 Historical studies, for example, have shown the origins of positivist epistemology to be a response to the ways in which the Reformation and the religious wars of the fifteenth and sixteenth centuries destroyed the foundations of certainty, dictated up to that time by the church. For those who believed that humankind could not live well without the existence of fixed categories of natural and social life — categories that impose themselves on everybody because of their undeniable validity — this collapse of authority was a primary concern (Wagner 1995).

6 In his book *The Tragedy of Political Science*, Ricci (1984, 296) points out that the classical study of politics was dominated by such normatively laden concepts of "authority," "justice," "patriotism," "responsibility," "virtue," "rights," "tyranny," and "nation." The tragedy of contemporary political science, in his view, is to be found in the fact that such concepts have in large part been replaced by quantitatively ori-

ented terms such as "system," "attitudes," "socialization," and "cognition." It is this shift in focus from the normative foundations of politics to the quantitatively operationalizable that has created the crisis of the discipline. Political scientists, in the process, have turned away from the kinds of questions that the members of society take to be important and relevant. In short, political science has sacrificed its relevance on the altar of statistical generalization.

7 The job today, according to Toulmin (1990, 193) is not to construct more comprehensive, timeless theories but rather "to limit the scope of even the best-framed theories, and fight the intellectual reductionisms that became entrenched during the ascendency of rationalism." This means a new form of knowing in which a much wider variety of multidisciplinary methods is brought into a more inclusive form of interdisciplinary reason.

8 Scientific progress, in Kuhn's view, involves a Darwinian competition among paradigms for superiority in problem-solving prowess. This guarantees scientific change but does not guarantee that science inevitably moves closer to the "truth."

9 After showing that positivism and its subsequent attempts to conclusively demarcate science and nonscience have failed, Laudan (1984) argues that the production of reliable knowledge is the distinguishing characteristic of any science. By this definition, "science" includes physics as well as military strategy, literary criticism, and the various policy sciences.

10 Postpositivism, Hawkesworth (1988, 191) explains, offers "policy analysis an alternative epistemology, a revised rhetoric, a reoriented methodology [and] . . . a different role for policy analysts in the political process. The fundamental task of theoretically informed policy analysis is to identify the dimensions of contention surrounding policy questions. Examining the conceptual and methodological assumptions that structure the constitution of facticity, the generation of evidence, the development of policy arguments, and the identification of policy alternatives can illuminate the forces circumscribing policy choices." In the attempt "to illuminate the political dimensions of perception and cognition, the influence of theoretical assumptions upon choice and action, the contentious character of scientific policy prescriptions, postpositivist policy analysis involves a more participatory conception of democracy. . . . In illuminating the precise grounds upon which specific decisions are made, post-positivist policy analysis can facilitate awareness of the character of the world which is being shaped and of viable alternatives. Rather than encouraging resignation to fate or blind sub-

mission to the status quo, post-positivist policy analysis can contribute to the choice of a way of life" (Hawkesworth, 1988, 192–93).

11 With regard to this relationship, Innes (1990, 20) writes: "as data enter the policy language, they become part of problem definitions, they set boundaries on possibilities for solutions, and they define the standards for choosing actions and evaluating results. Thus, the data affect policy not so much because of the facts that they reveal as because the concepts implicit in them become implicit in the discussion."

Chapter 5. Science and Politics in Environmental Regulation

1 For a discussion of this process in the case of setting clean air standards, see S. Melnick 1983.

2 The central focus in disputes about unknowns has centered on the concept of "margin of safety." As a result of the wide gap between what was known and what was unknown, some have argued that it would be unwise to permit exposures to rise to the level of known effects. In this view, a cushion should be maintained between known effects and allowable exposures. This margin has also been justified on the grounds that evolving knowledge demonstrates adverse effects at decreasing levels, that this trend would be expected to continue, and that it should be provided for in acceptable levels of exposure. But those who have justified the higher levels of exposure sought to set the limits only in terms of the fairly known effects. Given the controversies in the 1970s over environmental effects, this emphasis on margins of safety became increasingly problematic.

3 As we saw in chapter 4, basic to this research is an effort to determine what makes scientists accept some claims as better than others, given that such determination cannot be decided through a simple appeal to the external world. Historical accounts of the ascent and fall of particular scientific theories, ethnographic investigations of laboratory work, and examination of public controversies pertaining to science and technology have all supplied important insights into the processes by which an interpretation of reality obtains acceptance as the real thing.

Chapter 6. Confronting Experts in the Public Sphere

1 The term "environmental movement" tends to obscure the fact that there has been a diversity of movements within the environmental

camp. In the interests of clarity and simplicity, we use the term here to refer to the more left-leaning segments of the movements, in particular those critical of capitalism and the ideologies of economic growth.

2 The politics of the environmental movement portray science as an obstacle to the expression of environmental concerns. Typically, science is said to be used to silence concerns about the world in which we live rather than to enable and empower those concerns. Fears about the environment are met with scientifically based reassurances that all is well, even though citizens' experiences may suggest the opposite. Risk assessment, as we saw in the previous chapter, was seen to be introduced for this reason.

3 Writers such as Gurr (1985) and Kaplan (2000) emphasize the role of conflicts over the uses of scarce resources in the twenty-first century. For Gurr, the next century will see the rise of a new militaristic authoritarianism in the struggle over access to the globe's limited resources. It is, in fact, a goal for which the State Department and Pentagon already make contingency plans under the title of "environmental security." Such political forces are seen to only enhance the trends toward more centralized, technocratic government.

Chapter 7. Not in My Backyard

1 Portney (1991, 11) gives the following example: "Nearly everyone seems to agree that more prison space is needed if the criminal justice system is to be able to treat convicted criminals as harshly as the public mood warrants. Yet no one wants a prison in his or her city or town. . . . Most people seem to agree that such facilities are a necessary and acceptable result of living in an industrial society." In more recent years, it should be pointed out, some communities have actually sought out such facilities. It has been accepted as a solution to deteriorating economic conditions.

2 In point of fact, the situation is more complicated. Studies show that the risk involved in driving to the airport correlates with the age of the driver. Statistically speaking, for young drivers, it is safer to fly in the plane. This is not necessarily the case for older drivers (Lopes 1987).

3 There have been no major sitings of hazardous waste incinerators in the United States since the late 1970s.

4 Some have argued that the Swan Hills case is biased because the residents live in a stagnating economic area and could have been influenced by the economic incentives to participate rather than moved by

the force of the better argument. But Barry Rabe, the expert on the subject of the Swan Hills case, disagrees. He writes the following: "Some analysts dismiss the Swan Hills case, contending that a combination of its economic status and unusually isolated location contributed to a siting success that is unlikely to recur. . . . These arguments have some merit but cannot be taken too seriously. Numerous other communities in Alberta, as well as other provinces and states, feature many of the same qualities as Swan Hills" (Rabe 1994, 87).

5 In a case study of the siting of a hazardous waste facility in Minnesota, McAvoy (1999) provides important evidence of what he calls "citizen rationality" in a case of NIMBY. In his critique of technocracy, however, he neglects to differentiate sufficiently between state officials and their technical experts. Tesh (1999) also provides evidence of the abilities of ordinary citizens to deal meaningfully with technical issues. Unfortunately, she tends to see this as a challenge to the argument that citizens rely on social considerations such as trust. From the perspective advanced in this chapter, this view would seem to depend on a traditional understanding of science. Emphasis on social considerations, in the constructivist view, is inherent to scientific expertise, whether on the part of the scientific expert or the citizen expert.

6 With regard to the industrial perspective, one only needs to read the running commentaries against environmentalism that Mobil Oil publishes regularly on the Op-Ed page of the *New York Times*.

Chapter 8. Citizens as Local Experts

1 I am indebted to Steven Sperling, research director of the Citizen's Clearinghouse for Hazardous Waste, for discussing with me the nature of the organization's technical assistance to local communities.

2 The following account of Woburn is drawn from the works of P. Brown and Mikkelsen (1990).

3 For a fascinating story of the legal action against these companies, see the best-seller *A Civil Action*, also a Hollywood motion picture.

4 I would like to thank Dr. John Kurien and Dr. K. J. Joseph of the Center for Development Studies in Trivandrum, N. C. Narayanan of the Institute for Social Studies in The Hague, Babu Ambat of the Centre for Environment and Development in Trivandrum, and Richard Franke of Montclair State University for their helpful thoughts and comments on both the People's Planning Campaign and participatory resource mapping.

5   Since the 1950s, India has regularly engaged in a planning process. Impressed with the idea that a Soviet-style model of five-year planning could help to speed along the struggle to modernize, Nehru mandated a system of five-year planning processes that is still in place today. Much like the experience elsewhere, however, centralized planning has proven a disappointment. In response, the central government passed a number of constitutional amendments designed to facilitate the devolution of the process to the local level and established a planning committee to explore and monitor the development of decentralization. Despite the acclaimed benefits of the decentralization of planning, as well as repeated commitments to it, the planning process in India has remained highly centralized and bureaucratic. There have been few serious efforts at the state level to develop the process.

The major reason for the failure of these planning efforts has been the absence of popular representative administrative structures below the state level. Local institutions have seldom been given the power or financial resources to enable serious development interventions. Furthermore, New Delhi has also contributed to the failure by refusing to devolve more powers to the states while continuing to thrust on them one new centrally sponsored program after another.

The failure led to a national study commission on decentralization, which gave rise to a national debate and several constitutional amendments that officially empowered the local level with a mandated role in the planning process. Nonetheless, the general experience following these provisions has been not encouraging. In many cases, the mandates have simply been ignored; in others, the machinery and resources to make it possible still do not exist at the local level.

6   Within ten years, KSSP had established three magazines, one for academic interest, one for children, and one for high school students. Still in publication, they sell thousands of copies each year, approximately 100,000 copies combined.

7   Not only was this considered a major environmental victory in Indian environmental struggles, but the campaign, according to R. Radhakrishnan, president of KSSP, gave birth to the term "sustainable development," later to be picked up in the international arena.

8   A lakh is a unit of measurement unique to India. One lakh equals 100,000.

9   Kerala, as a result, became the most literate state in India. Such literacy has turned an otherwise uneducated peasant population into a state of alert newspaper readers who have a sense of their own social and political interests.

10 The "alternative Nobel Prize" is formally known as the Right Livelihood Award. It has been granted each year since 1989 by the Right Livelihood Award Foundation in Sweden, established in 1980. The prize honors and supports those offering practical and exemplary answers to the crucial problems facing the world today.

11 On the technique of participatory rural appraisal, see Chambers 1997.

12 I owe these comments to R. Radhakrishnan, the president of KSSP, and to Dr. Babu Ambat, director of the Centre for Environment and Development in Trivandrum.

13 My discussion of participatory resource mapping is based in significant part on an interview with Dr. Ajaykumar Varma, principal scientific officer of the Science, Technology, and Environment Department of the state of Kerala.

14 In the other cases, the planning board has, in the interests of time and efficacy, encouraged the panchayats to use the less rigorous but quicker method of rapid rural appraisal, in particular "transect walks."

Chapter 9. Community Inquiry and Local Knowledge

1 The Highlander Center is a sixty-seven-year-old private, nonprofit, community-based popular education organization located in New Market, Tennessee. Originally founded on the citizen-oriented model of the Danish folk school, the center's programs are based on a nontraditional vision of adult education. Its educational programs challenge citizens to engage in political struggles to bring about social change in community life. Long involved with marginalized working peoples, disadvantaged community groups, and grassroots movements in poverty-ridden regions of Appalachia, the center has emphasized the methods of participatory research and community-based education in its work to fight economic injustice through democratic control of community life. Highlander's emphasis on participatory research has contributed both to a better understanding of how social scientists relate — and should relate — to the communities with which they interact and to more general efforts to rethink conventional educational methods. For its efforts in these directions, particularly as they have pertained to human rights, the center was nominated for a Nobel Prize in 1982. For an earlier history of the center, see Adams and Horton 1975.

2 For an overview of this emerging movement, see Sclove et al. 1998. Community-based research, according to their findings, is best de-

scribed as "research that is conducted by, with or for communities." The study locates approximately fifty community research centers in the United States, with the total number of projects conducted ranging annually from anywhere between 400 and 1,200. As the description suggests, these organizations vary significantly in their use of collaborative research methods. Some appear to engage in full-scale participatory research projects, assisting citizens in doing their own investigations. More typical, however, are projects that offer advice to community groups based on research conducted for local citizens, involving varying degrees of citizen consultation.

3 Other approaches included "co-operative inquiry" (Heron 1981), "action inquiry" (Torbert 1991), "participatory rural appraisal" (Chambers 1997), "applied anthropology" (Stull and Schensul 1987), "appreciative inquiry" (Cooperrider and Srivastava 1987), "action science" (Argyris et al. 1985), "research partnerships" (Whitaker et al. 1990), and "critical ethnography" (Quantz 1992), among others.

4 For present purposes, the discussion draws heavily on the publications and projects of a group called the "participatory research network." See, for example, Society for Participatory Research in Asia. Kassam and Mustafa 1982; and W. Fernandes and Tandon 1981. These publications contain bibliographies and discussions of case studies drawn from projects in the United States, Africa, Latin America, India, England, Canada, and Indonesia, as well as other information. In the United States, the Highlander Center in Tennessee is perhaps the most important ongoing institution engaged in participatory research (chap. 9, n. 1).

5 There are exceptions here. To add to the ambiguity of these distinctions, Fals-Borda and Rahman (1991), writers closely associated with the liberation tradition of participatory research, use the term "participatory action research."

6 Most researchers accept this distinction, although those holding a radical social constructivist or postmodern position reject it. This volume accepts the distinction as a necessary heuristic device, recognizing that any separation in the strict sense of the word is artificial and untenable.

7 Reason (1994, 329) writes: "Community meetings and events of various kinds are an important part of [participatory research], serving to identify issues, to reclaim a sense of community and emphasize the potential for liberation, to make sense of information collected, to reflect on progress of the project, and to develop the ability of the community to continue the [participatory research] and developmental process. These meetings engage in a variety of activities that are in

keeping with the culture of the community and might look out of place in an orthodox research project. Thus storytelling, socio-drama, plays and skits, puppets, song, drawing and painting, and other engaging activities encourage a social validation of 'objective' data that cannot be obtained through the orthodox processes of survey and field research. It is important for an oppressed group, which may be part of a culture of silence based on centuries of oppression, to find ways to tell and thus reclaim their own story."

8 Although it is not the job of the postpositivist policy analyst to proselytize, the task is more than just offering instrumental answers about how to efficiently achieve a given set of goals. Rather, the postpositivist analyst takes on the assignment of facilitating a dialogue among competing perspectives. In short, he or she is as much an educator as a substantive policy expert. The task is to help people see and tease out the assumptions and conflicts underlying particular policy positions, as well as the consequences of resolving them in one way or another. Rather than simply to supply answers, the job is to facilitate a dialogue that permits citizens to follow through a particular process that helps them arrive at their own answers. People confronted with the construction of an incinerator in their neighborhood, as we saw in chapters 7 and 8, should be assisted in making their own assessments of the arguments for and against on their own terms.

9 On training and training programs, see Society for Participatory Research in Asia 1987; Heron 1989; D'Abreo 1981; Bobo et al. 1991; and Torbert 1981.

Chapter 10. Ordinary Local Knowledge

1 One of the problems concerns the meaning of "local." Local knowledge can refer to knowledge about a specific local context, or it can more generally refer to all forms of knowledge. Postmodernists, for instance, emphasize that all knowledge originates in — and is thus influenced by — the local context in which that knowledge is generated or produced. In this view, as we saw in chapter 4, all knowledge — whether pertaining to the local or the global — is produced by, and grounded in, local practices. The emphasis in this work is on knowledge about local context, although the argument is sympathetic to the later position as well.

2 Although ordinary knowledge is corrigible, as Lindblom and Cohen readily concede, it nonetheless merits the term "knowledge." For one

thing, scientific findings are themselves fallible. Propositions that later prove to be false are at the outset considered to be knowledge. For another, knowledge, whether accurate or not, is knowledge to those who hold it to be the justification for specific actions, a point central to politics in general. Indeed, in politics one of the most fundamental questions is whose knowledge counts as knowledge. In the course of political conflict, one group's ordinary knowledge is pitted against another group's.

3 "The cognitive contributions of folk wisdom to technical knowledge," as Krimsky puts it, include "pragmatic knowledge obtained through the intergenerational transmission of trial-error experiences, intuitive understanding of complex interactive systems, the generation of scientific hypotheses, and causal links such as identification of the environmental sources of human disease or ecological degradation and an understanding of meaning and value of urban value" (Krimsky 1984, 253). With regard to the trial and error methods of earlier societies, generations of accumulated experiences typically led to accepted practices based on proven results.

4 The concept of "tacit knowledge" refers to knowledge that influences thought and behavior that is not ordinarily accessible to consciousness, but that under certain conditions or circumstances can be brought to awareness. In Polyani's 1983 view, much of one's competence is achieved through the tacit dimensions of human interaction. People know much more than they can express in words, and this unspoken tacit knowledge is an important aspect of their skills or competence.

5 "Recent archaeological evidence," as Kurien writes, "affirms that the Indian subcontinent had a maritime tradition dating back to the second and third millennium B.C. . . . What traditional fishermen and seamen shared in common was the science of navigation and the vast accumulated fund of knowledge about the sea. . . . The elaborate understanding of the nuances of the aquatic milieu and the behavior patterns of living marine organisms . . . are the quintessence of their knowledge system" (Kurien 1988, 476).

6 In Kurien's words, "Fishermen can rarely make explicit any general 'theory' of their fishing. We may infer that their 'theory' is constructed from observation and tested by further observation. They add or subtract from 'theory' by producing new explanations or dropping existing ones. The process defies verbalisation in the form of general axioms on the practice of fishing. . . . Compared to the intricate knowledge of the totality of the eco-system [in taxonomy and biology . . . fishermen have a rather simple . . . taxonomy and biology of marine organisms.

They point out that such specialized knowledge of biology is of little practical use to them. It hardly augments their ability to catch fish" (Kurien 1988, 477). The same fishermen also exhibit "a conservational ethics toward fishery resources" grounded in "the view of 'mother ocean' as a life-giving system rather than a hunting ground."

7 With the arrival of political independence, as Kurien (1988, 477–78) explains, "the proponents of planned development . . . stressed the need for modernisation . . . to achieve higher levels . . . of production." As a result, "a process of . . . bureaucratization of fishery science and technology was undertaken with fervor. . . . A series of central institutes with research stations all over the country began the . . . task of systematizing data collection on fish." This entailed a steady substitution of indigenous institutions and cultural knowledge with formal institutions and objectified knowledge. "The new fishery science emphasized a taxonomical approach. The terminology and methodology adopted by Indian fishery scientists is understood by other scientists in the West, facilitating easy communications between them" lending credence to the view "that knowledge is universally valid, objective and politically neutral. . . . This reductionist approach of studying individual species of fish does lead to a study of the nature process in the aquatic milieu. However, this approach is limited . . . [in that it] is compartmentalized and by no means aggregates to a *holistic* understanding of the ecosystem. It can be caricaturized as the 'fish-eye view of the sea' — valid, but certainly not a picture of totality."

8 From 1961 to 1981 in Kerala there were impressive increases in the harvest of fish, along with productivity increases. Thereafter, there were sharp decreases in both measures. Commercial fishing exports, however, showed regular increases during the period as a whole. The experts in the fishery institutions, in Kurien's words, did "not see any cause for anxiety in respect of the fishery resource in Kerala State. Their argument is that there is no 'biological overfishing'" (Kurien 1988, 479). On the other hand, the fishermen maintain "that the basis of the resource crisis is not so much the danger of the 'running out' of the resource or reaching its physical limit, but rather the more fundamental issue of possible disruption of the ecological support systems in the ocean which indeed sustain the stock."

9 The search for active ingredients in indigenous medicines by pharmacologists, biochemists, and ethnobotanists has led to the development of important modern medicines (Brush and Stabinsky 1996). From ethnomedicine, for example, we have learned that the American Indians had discovered herbs that worked as oral contraceptives, and

that a plant found in the West Indies was used by indigenous peoples of the islands to successfully treat Hodgkin's disease and forms of leukemia. More than 150 drugs used by North American Indians have subsequently played a role in modern pharmacology in the United States. In addition to medicine, a greater understanding of earlier agricultural methods and tropical forests has also supplied important insights and knowledge about the nature of ecological systems. To recognize such indigenous knowledge is to draw on thousands of years of practical experience. In some cases, research into these forms of knowledges has led to important hypotheses, the testing of which has led to important discoveries. In other cases, such investigation discovered the information directly.

10 Irwin (1995, 18) describes a case involving a highly public dispute in 1980 between the National Union of Agricultural Workers and Allied Workers and the British regulatory authorities over the herbicide, 2, 4, 5-T. By the time of the dispute, "2, 4, 5-T had already been controversial for some time because of its allegedly hazardous properties (chloracne, birth defects, spontaneous abortion, cancer) and also for its overall impact on the natural environment." In view of international attention given to these hazards, "a number of countries had at that time either banned or severely restricted the use of the herbicide . . . [but] the British regulatory authorities had historically been resistant to the ban."

11 The farmworkers union presented the Advisory Committee on Pesticides (ACP) with a detailed "dossier" on the chemical. The union attempted to organize its own database by requesting survey information on the health effects of 2, 4, 5-T directly from its membership. The responses were then put together not in statistical form but as a series of case studies for submission to the advisory committee. For the farmworkers, this was a reasonable effort to synthesize information in a persuasive fashion. In their report, the farmers presented "cases where health damage is allegedly linked to 2, 4, 5-T exposure. . . . The overall conclusion of the farmworker submission was that: Considering the additional evidence which has not been evaluated by the ACP . . . it becomes absolutely incomprehensible that workers, their families and the general public can remain subject to the risks for one minute longer." In their response, "the ACP argued forcefully that 'there are no grounds to suggest a causal relationship with the stated effects' " (Irwin 1995, 19–20). It is no great surprise, however, that this conclusion did not change the view of the farmworkers.

12 The farmers in the campaign ridiculed the ACP's concept of "recom-

mended" working conditions, pointing to specific instances where such conditions were clearly violated. Specifically, they argued that such breaches were more than just periodic lapses — they were instead inevitable consequences of risky practices. To draw on one account, users, too, "are often simply unaware of the directions for use or, if they are aware of them, find that they are working under so much pressure that it is easier to ignore them." The conditions under which they find themselves are "all a long way from the laboratory conditions in which tests may be conducted." In particular, the farmworkers had knowledge of probable spraying conditions (high winds, thick undergrowth, hot weather, etc.). Pointing to a variety of operating circumstances (in terms of levels of work information about the chemicals employed, nonfunctioning or inadequate equipment, long distances from washrooms, inadequate facilities for the cleaning and disposal of containers, etc.), the farmworkers showed that the ACP's concept of "normal operation" was seriously flawed.

13 Such research emerged in significant part with the so-called green revolution that was exported by the advanced industrial countries in the 1960s to the developing world. The goal was to increase crop yields through the introduction of hybrid seed strains.

14 Such local knowledge, as Bourdieu (1977) puts it, is a kind of knowledge that does not move through a theoretical state in which a "discourse" develops. Rather, it goes directly from practice to practice.

15 Unfortunately the journal has fallen victim to hard financial times and is no longer being published. There have been discussions to get it going again, but as of yet, this has not happened. For this information I thank Dick Cloete, Johannesburg environmental activist and former editor of *New Ground*.

16 Liebenburg, in fact, seeks to establish a connection between tracking and the early development of science. He argues, in this respect, that the original speculative hypotheses of early hunter-gathers bear a direct relationship to the theoretical propositions of modern physics in its effort to "track" submolecular particles. The art of tracking, he maintains, constitutes an early attempt to transcend the boundaries of direct empirical observation through the use of intuitive and speculative thought processes.

17 There is, moreover, no alternative to such ordinary knowledge. Not only is most of the basic information necessary for political action in the form of ordinary knowledge, but policy science has offered us nothing with which we might replace it. At the most basic level, our store of ordinary knowledge provides us with the basic information —

for example, different levels of government have responsibility for passing particular laws, administrative agencies carry out specific functions, poor people can apply for social welfare, some policemen accept bribes, airline competition tends to lower ticket prices, workers will go on strike if not paid enough money, or the word "felony" refers to specific categories of crime. For the citizen — the social scientist included — the most fundamental categories of knowledge employed in politics and social problem solving are ordinary. Such knowledge is the common denominator that makes society possible. Lindblom (1990), like John Dewey in this respect, conceives of scientific thinking as only a more rigorous and refined variant of the basic logic of ordinary thinking.

18 As Jacobs (1961, 441) writes, "The processes that occur in cities are not arcane, capable of being understood only by experts. They can be understood almost by anybody. Many ordinary people already understand them; they simply have not given the processes names, or considered that by understanding these ordinary arrangements of cause and effect, we can also direct them if we want to."

19 In countering the abstract models of empirical planning analysis, Jacobs (1961, 443) interestingly stresses what she calls the "pinpoint clues" of "unaveraged events." In her words, "This awareness of 'unaverage' clues — or awareness of their lack — is . . . something that any citizen can practice. City dwellers, indeed, are commonly great informal experts in precisely this subject. Ordinary people in cities have an awareness of 'unaverage' qualities which is quite consonant with the importance of these relatively small quantities." Here, she argues, the planners are at a disadvantage. Having been trained to discount the "unaverage quantities" as relatively inconsequential, they easily overlook that which is vital.

Chapter 11. Discursive Institutions for Environmental Policy Making

1 In this regard, Innes (personal communication, October 1998) explains that emphasis "on discourse does not eliminate the tendency for some participants to prefer other issues or to choose in self-interested ways," but it does "provide a counter-balance." Even though "some may argue that the way one talks is little more than rhetoric covering real motivations, this is an overly simplistic perspective. Forcing par-

ticipants to conduct a particular kind of discourse can lead many to internalize, or at least accept, values reflected in that discourse."

2   In addition to Berger's two books on the inquiry process, the story of the northern pipeline inquiry is presented in a film titled *The Inquiry*. The documentary follows the Berger Commission from beginning to end.

3   I owe a good deal of my understanding of the details of the consensus conference, as practiced in Denmark and elsewhere, to discussions with Simon Joss, a leading observer of the practice.

4   Many citizen panels employ random sampling in their selection of participants. In the case of the Plannungszelle, for example, all citizens age eighteen or older in a particular community or group have an equal chance of being selected (Dienel 1992).

5   In a consensus conference in England concerned with irradiation of food, for example, participants allotted only 20 percent of their final report to irradiation. Concluding that the process is unnecessary, they devoted the remaining 80 percent of the report to its alternatives.

Chapter 12. The Environments of Argument

1   The development of such practices might follow the lead of Technical Assistance Grants (TAG) offered by the Superfund program designed to deal with the cleanup of hazardous waste sites (Chess et al. 1990). Although TAGs are an important step in the right direction, the practice of supplying such assistance continues to rest on a traditional understanding of the top-down professional-client relationship.

2   Chambers (1981) offers interesting examples of such issues in communication from his experiences in rural development. As he puts it, "The most difficult thing for an educated expert to accept is that poor farmers may often understand their situations better than he does. . . . A medical doctor, an agronomist, an engineer, and economist and a sociologist, visiting the same village will see and inquire about very different things. They will gain very different, and partial, views of a whole that is seen differently, and more holistically, by the villagers themselves.

3   Here we can also gain important insights from the work by Mary Douglas (1986) on "how institutions think." Similarly, Innes (1990, 20) writes that "social institutions — whether they are formal organizations like the Department of State, institutionalized processes like the

public hearings on environmental impact, or social indicators that are accepted representations of particular concepts — encode and organize information." At times, such organizations "substitute for individualized decision making because they offer routine procedures, explanations, and norms." In this way, "institutions influence individual cognition and collective understanding. . . . " They help individuals "decide what is predictable and accepted and what is deviant and should therefore be given attention."

References

Aaron, H. J. 1978. *Politics and the Professors*. Washington, D.C.: Brookings Institution.

Adams, F., and M. Horton. 1975. *Unearthing Seeds of Fire: The Idea of Highlander*. Winston-Salem, N.C.: John F. Blair.

Adams, J. 1995. *Risk*. London: UCL Press.

American Talk Issues Foundation. 1994. "Steps for Democracy: The Many versus the Few." 24 (9–19 January).

Amy, D. 1987. "Can Policy Analysis Be Ethical?" In *Confronting Values in Policy Analysis*, ed. F. Fischer and J. Forester. Newbury Park, Calif.: Sage.

Archer, M. 1990. "Theory, Culture, and Post-industrial Society." In *Global Culture*, ed. M. Featherstone, 97–119. London: Sage Publications.

Argyris, C., et al. 1985. *Action Science*. Cambridge: Harvard University Press.

Ascher, M., and R. Ascher. 1981. *Code of the Quipu: A Study in Media, Mathematics, and Culture*. Ann Arbor: University of Michigan Press.

Atran, S. 1987. "Origins of the Species and Genus Concepts: An Anthropological Perspective." *Journal of History of Biology* 20:195–279.

Bachrach, P., and M. Baritz. 1961. "Two Faces of Power." *American Political Science Review*. 56: 947–52.

Bahro, R. 1987. *Logik der Rettung: Wer kann die Apokalypse aufhalten?* Stuttgart: Weitbrecht.

Baily, R. 1993. *Ecoscam: The False Prophets of Ecological Apocalypse*. New York: St. Martins's Press.

Barber, B. R. 1984. *Strong Democracy*. Berkeley: University of California Press.

References

Barker, B., and B. G. Peters, eds. 1993. *The Politics of Expert Advice.* Pittsburgh, Pa.: University of Pittsburgh Press.

Barnes, B., and D. Edge, eds. 1992. *Science in Context.* London: Open University Press.

Bast, J. L., P. J. Hill, and R. Rue. 1994. *Eco-Sanity: A Common-Sense Guide to Environmentalism.* Lanham, Md.: Madison Books.

Baudrillard, J. 1983. *In the Shadows of the Silent Majorities.* New York: Semiotext.

Baughman, M. 1995. "Mediation." In *Fairness and Competence in Citizen Participation: Evaluating Models for Environmental Discourse,* ed. O. Renn, T. Webler, and P. Wiedemann, 253–60. Dordrecht: Kluwer Academic Publishers.

Baumol, W. J. 1991. "Toward a Newer Economics: The Future Lies Ahead." *Economic Journal* 101, no. 1: 1–8.

Beck, U. 1986. *Risikogesellschaft: Auf dem Weg in eine andere Moderne.* Frankfurt am Main: Suhrkamp.

———. 1992. *Risk Society: Towards a New Modernity.* Newbury Park, Calif.: Sage Publications.

Beck, U., et al. 1994. *Reflexive Modernization.* Newbury Park, Calif.: Sage.

———. 1995a. *Ecological Enlightenment: Essays on the Politics of the Risk Society.* Atlantic Highlands, N.J.: Humanities Press International.

———. 1995b. *Ecological Politics in an Age of Risk.* London: Polity Press.

Bell, D. 1971. "Technocracy and Politics." *Survey* 16: 10.

———. "The Social Framework of the Information Society." In *The Computer Age: A Twenty Year View,* ed. M. Dertouzos and J. Moses. Cambridge: MIT Press.

Bellini, James. 1989. *High Tech Holocaust.* San Francisco: Sierra Club Books.

Bendiger, J. R. 1986. *The Control Revolution: Technological and Economic Origins of the Information Society.* Cambridge: Harvard University Press.

Beneveniste, G. 1987. "Some Functions and Dysfunctions of Using Professional Elites in Public Policy." In *Research in Public Policy Analysis and Management,* ed. Stuart Nagel. Vol. 3. Greenwich, Conn.: JAI Press.

Bennahum, D. S. 1995. "Mr. Gingrich's Cyber-Revolution." *New York Times,* January 17, A19.

Bennett, D. 1986. "Democracy and Public Policy Analysis." In *Research in Public Policy Analysis and Management,* ed. Stuart Nagel. Vol. 3. Greenwich, Conn.: JAI Press.

Bennett, L. W. 1992. *The Governing Crisis: Media, Money, and Marketing in American Elections.* New York: St. Martin's Press.

Bennis, W. 1966. *Changing Organizations*. New York: McGraw-Hill.

Berger, T. 1977. *Northern Frontier, Northern Homeland: The Report of the Mackenzie Valley Pipeline Inquiry*. Vols. 1–2. Ottawa: Supply and Services.

———. 1985. *Village Journey: The Report of the Alaska Native Review Commission*. New York: Wang and Hill.

Berlin, B. 1992. *Ethnobiological Classifications: Principles of Categorization of Plants and Animals in Traditional Societies*. Princeton, N.J.: Princeton University Press.

Berlin, B., D. Breedlove, and P. Raven. 1973. "General Principles of Classification and Nomenclature in Folk Biology." *American Anthropologist* 75: 214–42.

Bernal, J. D. 1969. *Science in History*. Vol. 1. London: Pelican.

Bernstein, R. J. 1976. *The Restructuring of Social and Political Theory*. New York: Harcourt Brace and Jovanovich.

———. 1983. *Beyond Objectivism and Relativism: Science, Hermeneutics, and Praxis*. Philadelphia: University of Philadelphia Press.

Berten, H. 1995. *The Idea of Postmodernism: A History*. London: Routledge.

Berube, M. 1996. "Public Perceptions of the Universities and Faculty." *Academe* 82, no. 4 (July–August): 10–17.

Best, J., ed. 1989. *Images of Issues*. New York: Aldine de Gruyter.

Blaug, R. 1991. *Democracy Real and Ideal: Discourse Ethics and Radical Politics*. Albany: State University of New York Press.

Blinder, A. 1997. "Is Government Too Political?" *Foreign Affairs* 76, no. 6: 115–26.

Bobbio, N. 1987. *The Future of Democracy: A Defense of the Rules of the Game*. Cambridge: Polity Press.

Bobo, K., J. Kendall, and S. Max. 1991. *Organizing for Social Change*. New Market, Tenn.: Highlander Center.

Bobrow, D., and J. Dryzek. 1987. *Policy Analysis by Design*. Pittsburgh, Pa.: University of Pittsburgh Press.

Bohman, J. 1996. *Public Deliberation: Pluralism, Complexity, and Democracy*. Cambridge: MIT Press.

Bohman, J., and W. Rehg, eds. 1997. *Deliberative Democracy*. Cambridge: MIT Press.

Boloch, B., and H. Lyons. 1993. *Apocalypse Not: Science, Economics, and Environmentalism*. Washington, D.C.: Cato Institute.

Bookchin, M. *The Ecology of Freedom*. Palo Alto, Calif.: Cheshire.

Bourdieu, P. 1977a. *Outline of a Theory of Practice*. London: Cambridge University Press.

———. 1977b. "Symbolic Power." *Critique of Anthropology* 13–14 (summer): 77–85.

Bowers, C. A. 1982. "The Reproduction of Technological Consciousness: Locating the Ideological Foundations of a Radical Pedagogy." *Teachers College Record* 83, no. 4 (summer 1982): 531.

Breyman, S. 1993. "Knowledge as Power: Ecology Movements and Global Environmental Problems." In *The State and Social Power in Global Environmental Politics*, ed. R. D. Lipschutz and K. Conca, 124–57. New York: Columbia University Press.

———. 1998. *Movement Genesis: Social Movement Theory and the West German Peace Movement*. Boulder, Colo.: Westview Press.

Brint, S. 1994. *In an Age of Experts*. Princeton, N.J.: Princeton University Press.

Brokensha, D., M. Warren, and O. Werner, eds. 1980. *Indigenous Knowledge Systems and Development*. Lanham, Md.: University of America Press.

Brookfield, S. D. 1986. *Understanding and Facilitating Adult Learning*. San Francisco: Jossey-Bass.

Brooks, H. 1965. "Scientific Concepts and Cultural Change." *Daedalus* 94 (winter): 68.

Brown, N. 1977. *Perception, Theory, and Commitment: The New Philosophy of Science*. Chicago: Precedent Publishing.

Brown, P. 1990. "Popular Epidemiology: Community Response to Toxic Waste–Induced Disease." In *The Sociology of Health and Illness in Critical Perspective*, ed. P. Conrad and R. Kern, 77–85. New York: St. Martin's Press.

Brown, P., and E. J. Mikkelsen. 1990. *No Safe Place: Toxic Waste, Leukemia, and Community Action*. Berkeley: University of California Press.

Brush, S. B. 1996. "Whose Knowledge, Whose Genes, Whose Rights?" In *Valuing Local Knowledge: Indigenous People and Intellectual Rights*, ed. S. B. Brush and D. Stabinsky, 1–31. Washington, D.C.: Island Press.

Brush, S. B., J. C. Heath, and Z. Huaman. 1981. "Dynamics of Andean Potato Agriculture." *Economic Botany* 35, no. 1: 70–88.

Brush, S. B., and S. Stabinsky. 1996. *Valuing Local Knowledge: Indigenous People and Intellectual Property Rights*. Washington, D.C.: Island Press.

Buergergutachten Uestra: Attraktiver Oeffentlicher Personennahverkehr in Hannover. 1996. Bonn: Stiftung Mitarbeit.

Bullard, R. D. 1993. *Environmental Justice and Communities of Color*. San Francisco: Sierra Club.

Busenberg, G. J. 1999. "Collaborative and Adversarial Analysis in Environmental Policy." *Policy Sciences* 32, no. 1: 1–11.

Caldwell, L. K. 1975. "Managing the Transition to Post-modern Society." *Public Administration Review* 35, no. 6: 567–72.

Cambrosio, A., and Limoges, C. 1991. "Controversies as Governing Processes in Technology Assessment." *Technology Analysis and Strategic Management* 3: 337–95.

Cancian, F. M., and C. Armstead. 1992. "Participatory Research." In *Encyclopedia of Sociology*, ed. E. F. Borgatta and Maria L. Borgatta, 1427–32. New York: Macmillan.

Carson, R. 1962. *Silent Spring*. Boston: Houghton Mifflin.

Castells, M. 1996. *The Rise of the Networked Society*. London: Blackwell.

Chambers, R. 1981. "A Lesson for Rural Developers: The Small Farmer as Professional." *Development Digest* 19, no. 3: 3–12.

———. 1997. *Whose Reality Counts? Putting the First Last*. London: Intermediate Technology Publications.

Chess, C., and P. M. Sandman. 1989. "Community Use of Quantitative Risk Assessment." *Science for the People* (January–February): 20.

Chess, C., S. K. Long, and P. M. Sandman. 1990. *Making Technical Assistance Grants Work*. New Brunswick, N.J.: Environmental Communications Program.

Chomsky, N. 1989. Interview by B. Moyers. *A World of Ideas*, ed. B. Moyers. New York: Doubleday.

Churchman, C. W. 1971. *The Designing of Inquiring Systems*. New York: Basic Books.

Cohn, J. 1999. "Irrational Exuberance: When Did Political Science Forget About Politics?" *The New Republic Online* (October 25): 1–14.

Collette, W. 1987. *How to Deal with a Proposed Facility*. Arlington, Va.: Citizen's Clearinghouse for Hazardous Wastes.

Collette, W., and L. M. Gibbs. 1985. *Experts: A User's Guide*. Arlington, Va.: Citizen's Clearinghouse for Hazardous Wastes.

Collingsridge, D., and C. Reeves. 1986. *Science Speaks to Power: The Role of Experts in Policymaking*. New York: St. Martin's Press.

Collins, H. 1992. *Changing Order: Replication and Induction in Scientific Practice*. Chicago: University of Chicago Press.

Comstock, D. E., and R. Fox. 1993. "Participatory Research as Critical Theory: The North Bonneville, USA Experience." In *Voices of Change: Participatory Research in the United States and Canada*, ed. P. Park et al., 103 24. Toronto: OISE Press.

Comte, A. 1830. *Cours de philosophie positive*. Vol. 6. Paris: Bachelier.

Conover, P. 1984. "How People Organize the Political World." *American Journal of Political Science* 28: 95–126.

References

Conover, P., S. Feldman, and A. Miller. 1991. "Where Is the Schema?" *American Political Science Review* 85: 1357–80.

Cooperrider, D. L., and Srivastava, S. 1987. "Appreciative Inquiry in Organizational Life." In *Research in Organizational Change and Development*, ed. W. Pasmore and R. Woodman, 129–69. Greenwich, Conn.: JAI.

Cornwall, A., and R. Jewkes. 1995. "What Is Participatory Research?" *Soc. Sci. Med.* 41, no. 12: 1667–76.

Covello, V. T. 1993. *Risk Assessment Methods*. New York: Plenum Press.

Cranor, C. F. 1993. *Regulating Toxic Substances: A Philosophy of Science and Law*. New York: Oxford University Press.

Crosby, N. 1995. "Citizens Juries: One Solution for Difficult Environmental Problems." In *Fairness and Competence in Citizen Participation: Evaluating Models for Environmental Discourse*, ed. O. Renn, T. Webler, and P. Wiedermann, 157–74. Dordrecht, Netherlands: Kluwer.

Crosby, N., J. Kelly, and P. Schaeffer. 1986. "Citizens Panels: A New Approach to Citizen Participation." *Public Administration Review* 46: 170.

Crouch, E., and R. Wilson. 1982. *Risk/Benefit Analysis*. Cambridge, Mass.: Ballinger.

Crowfoot, J. E., and J. M. Wollendeck. 1990. *Environmental Disputes: Community Involvement in Conflict Resolution*. Washington, D.C.: Island Press.

Crozier, M., et al., eds. 1975. *The Crisis of Democracy*. New York: Trilateral Commission.

D'Abreo, D. A. 1981. "Training for Participatory Evaluation." In *Participatory Research and Evaluation: Experiments in Research as a Process of Liberation*, ed. W. Fernandes and R. Tandon. New Delhi: Indian Social Institute.

Dahl, R. 1989. *Democracy and Its Critics*. New Haven, Conn.: Yale University Press.

Danziger, M. 1995. "Policy Analysis Postmodernized: Some Political and Pedagogical Ramifications." *Policy Studies Journal* 23, no. 3: 435–50.

Davidson, P. 1965. "Advocacy and Pluralism in Planning." *AIP Journal* (November): 331–38.

Dear, M. 1992. "Understanding and Overcoming the NIMBY Syndrome." *Journal of the American Planning Association* 58 (summer).

deLeon, P. 1988. *Advice and Consent: The Development of the Policy Sciences*. New York: Russell Sage Foundation.

———. 1992. "The Democratization of the Policy Sciences." *Public Administration Review* 52 (March–April): 125–29.

Deleuze, G., et al. 1987. *A Thousand Plateaus*. London: Athlone.

de Roux, G. I. 1991. "Together against the Computer: PAR and the Struggle of Afro-Colombians for Public Services." In *Action and Knowledge: Breaking the Monopoly with Participatory Action-Research*, ed. O. Fals-Borda and M. A. Rahman, 37–53. New York: Apex Press.

Dewey, J. 1927. *The Public and Its Problems*. New York: Swallow.

Dienel, P. C. 1989. "Contributing to Social Decision Methodology: Citizen Reports on Technological Projects." In *Social Decision Methodology for Technological Projects*, ed. C. Vlek and G. Cvetkovich, 133–52. Dordrecht, Netherlands: Kluwer Academic Publishers.

———. 1992. *Die Plannungszelle: Eine Alternative zur Establishment-Democratie*. Opladen: Westdeutscher Verlag.

Diesing, P. 1962. *Reason in Society: Five Types of Decisions in Their Social Context*. Urbana: University of Illinois Press.

Dionne, E. J., Jr. 1991. *Why Americans Hate Politics*. New York: Simon and Schuster.

DiPerna, P. 1985. *Cluster Mystery: Epidemic and the Children of Woburn*. St. Louis, Mo.: Mosby.

Doble, J., and A. Richardson. 1992. "You Don't Have to Be a Rocket Scientist." *Technology Review* (January): 51–54.

Douglas, M. 1986. *How Institutions Think*. New York: Syracuse University Press.

Douglas, M., and A. Wildavsky. 1982. *Risk and Culture: An Essay on the Selection of Technical and Environmental Dangers*. Berkeley: University of California Press.

Dowie, M. 1995. *Losing Ground: American Environmentalism at the Close of the Twentieth Century*. Cambridge: MIT Press.

Drucker, P. F. 1993. "The Rise of the Knowledge Society." *Wilson Quarterly* 17, no. 2 (spring): 52–71.

Dryzek, J. S. 1982. "Policy Analysis as a Hermeneutic Activity." *Policy Sciences* 14: 309–29.

———. 1990. *Discursive Democracy*. Cambridge: Cambridge University Press.

———. 1996. "Strategies of Ecological Democratization." In *Democracy and the Environment*, ed. W. M. Lafferty and J. Meadowcroft, 108–23. Cheltenham, U.K.: Edward Elgar.

———. 1997. *The Politics of the Earth*. Oxford: Oxford University Press.

Dunn, W. 1981. *Public Policy Analysis*. Englewood Cliffs, N.J.: Prentice-Hall.

Durning, D. 1993. "Participatory Policy Analysis in a Social Service Agency: A Case Study." *Journal of Policy Analysis and Management* 12, no. 2: 297–322.

References

Dutton, D. 1984. "The Impact of Public Participation in Biomedical Policy: Evidence from Four Case Studies." In *Citizen Participation in Science Policy*, ed. J. C. Peterson, 147–81. Amherst: University of Massachusetts Press.

Dyson, F. J. 1993. "Science in Trouble." *American Scholar* 62 (autumn): 513–25.

Eckersley, R. 1992. *Environmentalism and Political Theory*. Albany: State University of New York Press.

Edelman, M. 1964. *The Symbolic Uses of Politics*. Urbana: University of Illinois.

———. 1988. *Constructing the Political Spectacle*. Chicago: University of Chicago Press.

Edelstein, M. R. 1988. *Contaminated Communities*. Boulder, Colo.: Westview Press.

Eder, Klaus. 1996. "The Institutionalization of Environmentalism: Ecological Discourse and the Second Transformation of the Public Sphere." In *Environment and Modernity: Toward a New Ecology*, ed. S. Lash, B. Szerzynski, and B. Wynne Risk, 203–23. Newbury Park, Calif.: Sage.

Edmondson, R., and F. Nullmeier. 1997. "Knowledge, Rhetoric, and Political Action in Context." In *The Political Context of Collective Action: Power, Argumentation, and Democracy*, ed. R. Edmonson and F. Nullmeier, 210–38. London: Routledge.

Eldon, M. 1981. "Sharing the Research Work: Participative Research and Its Role Demands." In *Human Inquiry: A Sourcebook of New Paradigm Research*, ed. P. Reason and J. Rowan, 253–66. New York: John Wiley.

Eldon, M., and R. Chrisholm, eds. 1993. "Varieties of Action Research." *Human Relations*. Special edition 46, no. 2.

Elgin, C. Z. 1984. "Representation, Comprehension, and Competence." *Social Research* 51, no. 4: 905–26.

Elliot, M. L. P. 1984. "Improving Community Acceptance of Hazardous Waste Facilities through Alternative Systems of Mitigating and Managing Risk." *Hazardous Waste* 1: 397–410.

Ellul, J. 1964. *The Technological Society*. New York: Vintage.

Epstein, S. 1996. *Impure Science: AIDS, Activism, and the Politics of Knowledge*. Berkeley: University of California Press.

Erlandson, D. A., et al. 1993. *Doing Naturalistic Inquiry: A Guide to Methods*. Newbury Park, Calif.: Sage Publications.

Eulau, H. 1977. *Technology and Civility*. Stanford, Calif.: Hoover Institution.

Evans, S., and H. Boyte. 1992. *Free Spaces: The Sources of Democratic Change in America*. Chicago: University of Chicago Press.

306

Ezrahi, Y. 1990. *The Descent of Icarus: Science and the Transformation of Contemporary Democracy.* Cambridge: Harvard.

Falco, M. 1973. *Truth and Meaning in Political Science.* Columbus, Ohio: Merrill.

Fals-Borda, O., and M. A. Rahman. 1991. *Action and Knowledge: Breaking the Monopoly with Participatory Action Research.* New York: Apex Press, 1991.

Farrington, J., and A. Martin. 1987. *Farmer Participatory Research: A Review of Concepts and Practices.* ODI Agricultural Administration Network, discussion paper no. 19. London: Overseas Development Institute.

Fay, B. 1975. *Social Theory and Political Practice.* New York: Holmes and Meyers.

Feenberg, A. 1991. *Critical Theory of Technology.* New York: Oxford University Press.

Fernandes, W., and R. Tandon, eds. 1981. *Participatory Research and Evaluation: Experiments in Research as a Process of Liberation.* New Delhi: Indian Social Institute.

Fernandez, M. E. 1986. *Participatory-Action Research and the Farming Systems Approach with Highland Peasants.* SR-CRSP technical report no. 75. Columbia: University of Missouri Department of Rural Sociology.

Filmer, P., et al. 1973. *New Directions in Sociological Theory.* Cambridge: MIT Press.

Finer, H. 1941. "Administrative Responsibility and Democratic Government." *Public Administration Review* 1 (summer 1941): 335–50.

Fiorino, D. J. 1990. "Citizen Participation and Environmental Risk: A Survey of Institutional Mechanisms." *Science, Technology, and Human Values* 15: 226–43.

———. 1995. "Regulatory Negotiation as a Form of Public Participation." In *Fairness and Competence in Citizen Participation: Evaluating Models for Environmental Discourse,* ed. O. Renn, T. Webler, and P. Wiedemann. Dordrecht, Netherlands: Kluwer Academic Publishers.

Fischer, F. 1980. *Politics, Values, and Public Policy: The Problem of Methodology.* Boulder, Colo.: Westview Press.

———. 1990. *Technocracy and the Politics of Expertise.* Newbury Park, Calif.: Sage Publications.

———. 1991a. "American Think Tanks: Policy Elites and the Politicization of Expertise." Governance 4, no. 3 (July): 332–53.

———. 1991b. "Risk Assessment and Environmental Crisis: Toward an Integration of Science and Participation." *Industrial Crisis Quarterly* 5: 113–32.

———. 1992. "Participatory Expertise: Toward the Democratization of Policy Science." In *Advances in Policy Studies since 1950*, ed. W. Dunn and R. Kelly, 351–76. New Brunswick: Transaction Press.

———. 1993. "Policy Discourse and the Politics of Washington Think Tanks." In *The Argumentative Turn in Policy Analysis and Planning*, ed. F. Fischer and J. Forester. Durham, N.C.: Duke University Press.

———. 1995. *Evaluating Public Policy*. Chicago: Nelson-Hall.

———. 1995. "From Technocracy to Participatory Research: First World Practices and Third World Alternatives." In *Designers of Development: Intellectuals and Technocrats in the Third World*, ed. B. Galjart and P. Silva, 54–68. Leiden, Netherlands: CNWS Publications.

———. 1998. "Beyond Empiricism: Policy Inquiry in Postpositivist Perspective." *Policy Studies Journal* 26, no. 1: 129–47.

Fischer, F., and J. Forester. 1993. *The Argumentative Turn in Policy Analysis and Planning*. Durham, N.C.: Duke University Press.

Fischer, F., and J. Forester, eds. 1987. *Confronting Values in Policy Analysis: The Politics of Criteria*. Newbury Park, Calif.: Sage.

Fischer, F., and M. Hajer. 1999. *Living with Nature: Environmental Politics as Cultural Discourse*. Oxford: Oxford University Press.

Fishkin, J. S. 1996. *The Voice of the People: Public Opinion and Democracy*. New Haven, Conn.: Yale University Press.

Fiske, S. T., and S. E. Taylor. 1984. *Social Cognition*. Reading, Mass.: Addison-Wesley.

Forester, J. 1992. "Envisioning the Politics of Public-Sector Dispute Resolution." *Studies in Law, Politics, and Society* 12: 247–86.

———. 1999. *The Deliberative Practitioner: Encouraging Participatory Planning Processes*. Cambridge: MIT Press.

Forester, T., ed. 1985. *The Information Technology Revolution*. Cambridge: MIT Press.

Foster, J. 1980. "An Advocate Role Model for Policy Analysis." *Policy Studies Journal* 8, no. 6 (summer): 958–64.

Foucault, M. 1972. *The Archaeology of Knowledge*. New York: Pantheon.

———. 1973. *The Order of Things*. New York: Vintage Books.

———. 1977. *Discipline and Punish*. New York: Pantheon.

———. 1980. *Power/Knowledge: Selected Interviews and Other Writings*. New York: Pantheon.

———. 1983. "The Subject and Power." In *Michael Foucault: Beyond Structuralism and Hermeneutics*, ed. H. L. Dreyfus and Paul Rabinow. Chicago: University of Chicago Press.

———. 1984. *The Foucault Reader*. New York: Pantheon.

Franke, F. 1993. *Life Is a Little Better*. Boulder, Colo.: Westview Press.

Freire, P. 1970. *Pedagogy of the Oppressed*. New York: Seabury Press.

———. 1973. *Education for Critical Consciousness*. New York: Seabury Press.

Friedmann, J. 1973. *Retracking America*. New York: Anchor.

———. 1987. *Planning in the Public Domain*. Princeton, N.J.: Princeton University Press.

Friedrich, C. 1940. "Public Policy and the Nature of Administrative Responsibility." *Public Policy* 1: 3–24.

Fumento, M. 1993. *Science under Siege: Balancing Technology and the Environment*. New York: William Morrow.

Galbraith, J. K. 1967. *The New Industrial State*. Boston: Houghton Mifflin.

Galison, P. 1997. *Image and Logic: A Material Culture of Microphysics*. Chicago: University of Chicago Press.

Gastil, J. 1994. "Democratic Citizenship and the National Issues Forums." Ph.D. diss., University of Wisconsin–Madison.

Gaventa, J. 1980. *Power and Powerlessness: Quiescence and Rebellion in an Appalachian Valley*. Urbana: University of Illinois Press.

———. 1988. "Participatory Research in North America." *Convergence* 21: 19–29.

———. 1993. "The Powerful, the Powerless, and the Experts: Knowledge in the Information Age." In *Voices of Change: Participatory Research in the United States and Canada*, ed. P. Park et al., 20–46. Toronto: OISE Press.

Gaylin, W., et al. 1978. *Doing Good: The Limits of Benevolence*. New York: Pantheon.

Geertz, C. 1983. *Local Knowledge: Further Essays in Interpretive Knowledge*. New York: Basic Books.

Gerber, J. M. 1992. "Farmer Participation in Research: A Model for Adaptive Research and Education." *American Journal of Alternative Agriculture* 7, no. 3: 118–21.

Gibbs, L. M. 1982. *Love Canal: My Story*. Albany: University of New York Press.

———. 1986. "Health Surveys: Think Before You Count." *Everybody's Backyard* 3: 2–3.

Giddens, A. 1990. *Consequences of Modernity*. London: Polity Press.

———. 1995. *New Statesman and Society*, 7 April.

Gieryn, T. F. 1983. "Boundary-Work and the Demarcation of Science from Non science: Strains and Interests in Professional Interests of Scientists." *American Sociological Review* 48: 781–95.

Gladwin, C. H. 1989. "Indigenous Knowledge Systems, the Cognitive Revolution, and Agricultural Decision Making." *Agriculture and Human Values* 6, no. 3 (summer): 32–41.

References

Gladwin, T. 1970. *East Is a Big Bird: Navigation and Logic in Puluwat*. Cambridge: Harvard University Press.

Glasser, B. 1992. *Basics of Grounded Theory Analysis*. Mills Valley, Calif.: Sociology Press.

Gleick, J. 1987. *Chaos Theory: Making a New Science*. New York: Viking.

Goldblatt, D. 1996. *Social Theory and the Environment*. Cambridge: Polity Press.

Gottlieb, B., and A. Farquharson. 1985. "Blueprint for a Curriculum on Social Support." *Social Policy* (winter): 31–34.

Gramsci, A. 1971. *Selections from the Prison Notebooks*. New York: International Publishers.

Greenwood, D. J., and M. Levin. 1998. *Introduction to Action Research: Social Research for Social Change*. Newbury Park, Calif.: Sage.

Guba, E. G. 1990. *The Paradigm Dialog*. Newbury Park, Calif.: Sage Publications.

Gundersen, A. G. 1995. *The Environmental Promise: Of Democratic Deliberation*. Madison: University of Wisconsin Press.

Gurr, T. R. 1985. "On the Political Consequences of Scarcity and Economic Decline." *International Studies Quarterly* 29: 51–71.

Gusfield, J. 1981. *The Culture of Public Problems*. Chicago: University of Chicago Press.

Haas, P. 1992. "Epistemic Communities and International Policy Coordination." *International Organization* 46, no. 1 (winter): 1–35.

Habermas, J. 1970a. "On Systematically Distorted Communication." *Inquiry* 13: 205–18.

———. 1970b. *Toward a Rational Society*. Boston: Beacon Press.

———. 1973. *Legitimation Crisis*. Boston: Beacon Press.

———. 1987. *The Theory of Communicative Action*. Vol. 2, trans. T. McCarthy. Cambridge, Mass.: Polity.

Hadden, S. G. 1991. "Public Perception of Hazardous Waste." *Risk Analysis*, vol. 11, no. 1: 47–57.

Hajer, M. 1995. *The Politics of Environmental Discourse*. Oxford: Oxford University Press.

Halberstam, D. 1993. *The Best and the Brightest*. New York: Fawcett.

Hall, P. 1993. "Policy Paradigms, Social Learning, and the State." *Comparative Politics* 25: 275–96.

Hannigan, J. A. 1995. *Environmental Sociology: A Social Constructivist Perspective*. London: Routledge.

Haraway, D. 1991. *Simians, Cyborgs, and Women*. London: Free Press.

Hardin, G. 1968. "The Tragedy of the Commons." *Science* 162, 1,243–48.

Harding, S. 1986. *The Science Question in Feminism*. Ithaca, N.Y.: Cornell University Press.

Harmon, M. M., and R. Mayer. 1986. *Organization Theory for Public Administration*. Boston: Little, Brown.

Harr, J. 1996. *A Civil Action*. New York: Vintage Books.

Harvey, David. 1999. "The Environments of Justice." In *Living with Nature*, ed. F. Fischer and M. Hajer. Oxford: Oxford University Press.

Harwood, R. C. 1991. *Citizens and Politics: A View from Main Street America*. Dayton: Kettering Foundation.

Haskell, T. 1984. *The Authority of Experts*. Bloomington: Indiana University Press.

Hawkesworth, M. E. 1988. *Theoretical Issues in Policy Analysis*. Albany: SUNY Press.

Hays, S. 1987. *Beauty, Health, and Permanence: Environmental Politics in the United States, 1955–1985*. Cambridge: Cambridge University Press.

Healey, P. 1993. "Planning through Debate: The Communicative Turn in Planning Theory." In *The Argumentative Turn in Policy Analysis and Planning*, ed. F. Fischer and J. Forester, 233–53. Durham, N.C.: Duke University Press.

———. 1997. *Collaborative Planning*. London: Macmillan.

Heclo, H. 1974. *Modern Social Politics in Britain and Sweden*. New Haven, Conn.: Yale University Press.

———. 1976. "Conclusion: Policy Dynamics." In *The Dynamics of Public Policy*, ed. R. Rose. London: Sage.

———. 1978. "Issue Networks and the Executive Establishment." In *The New American Political System*, ed. A. King, 87–124. Washington, D.C.: American Enterprise Institute.

Heilbroner, R. 1974. *An Inquiry into the Human Prospect*. New York: Harper and Row.

Heineman, R. A., W. Bluhm, S. A. Peterson, and E. N. Kearney. 1990. *The World of the Policy Analyst: Rationality, Values, Politics*. Chatham, N.J.: Chatham House.

Heron, J. 1981. "Philosophical Basis for a New Paradigm." In *Human Inquiry: A Sourcebook of New Paradigm Research*, ed. P. Reason and J. Rowan. Chichester, U.K.: John Wiley.

———. 1989. *The Facilitator's Handbook*. London: Kogan Page.

———. 1992. *Feeling and Personhood: Psychology in Another Key*. London: Sage.

Hill, S. 1992. *Democratic Values and Technological Choices*. Stanford, Calif.: Stanford University Press.

References

Hirschhorn, L. 1979. "Alternative Service and the Crisis of the Professions." In *Co-ops, Communes, and Collectives: Experiments in Social Change in the 1960s and 1970s*, ed. J. Case and R. C. R. Taylor, 153–93. New York: Pantheon.

Hirschman, A. 1981. *Essays in Trespassing*. Cambridge, Mass.: Cambridge University Press.

Hiskes, A. L., and R. P. Hiskes. 1986. *Science, Technology, and Policy Decisions*. Boulder, Colo.: Westview Press.

Hiskes, R. P. 1998. *Democracy, Risk, and Community: Technological Hazards and the Evolution of Liberalism*. New York: Oxford University Press.

Hofferbert, R. I. 1990. *The Reach and Grasp of Policy Analysis*. Tuscaloosa: University of Alabama Press.

Hoffman, L. 1989. *The Politics of Knowledge: Activist Movements in Medicine and Planning*. Albany: State University of New York Press.

Hofmann, J. 1993. "Implicit Theories in Policy Discourse: Interpretations of Reality in German Technology Policy." *Policy Sciences* 18, no. 2: 127–48.

Hofrichter, R. 1993. "Cultural Activism and Environmental Justice." In *Toxic Struggles: The Theory and Practice of Environmental Justice*, ed. R. Hofrichter, 85–96. Philadelphia: New Society Publishers.

Holland, J., with J. Blackburn, eds. 1998. *Whose Voice? Participatory Research and Policy Change*. London: Intermediate Technology Publications.

Hoppe, R., and A. Peterse. 1993. *Handling Frozen Fire*. Boulder, Colo.: Westview Press.

Howard, A., Sir. 1924. *Co-production in India*. London: Oxford University Press.

Hudson, W. E. 1995. *American Democracy in Peril*. Chatham, N.J.: Chatham House Publishers.

Hughes, T. 1991. *The American Century*. Cambridge: Harvard University Press.

Illich, I. 1989. *A Celebration of Awareness: A Call for Institutional Revolution*. New York: Heyday Books.

Immerwahr, J., and J. Johnson. 1994. *Second Opinions: Americans' Changing Views on Healthcare Reform*. New York: Public Agenda Foundation.

Inglehart, R. 1971. *The Silent Revolution: Changing Values and Political Styles among Western Publics*. Princeton, N.J.: Princeton University Press.

Ingram, H. M., and S. Rathgeb Smith, eds. 1993. *Public Policy for Democracy*. Washington, D.C.: Brookings Institution.

Innes, J. 1990. *Knowledge and Public Policy.* 2d ed. New Brunswick, N.J.: Transaction Books.

———. 1998. "Information in Communicative Planning." *Journal of the American Planning Association* 64, no. 1: 52–63.

Irwin, A. 1995. *Citizen Science: A Study of People, Expertise, and Sustainable Development.* London: Routledge.

Irwin, A., and B. Wynne, eds. 1995. *Misunderstanding Science.* Cambridge: Cambridge University Press.

Jacobs, J. 1961. *The Death and Life of Great American Cities.* New York: Vintage Books.

Jaenicke, M. 1996. "The Environment and the Civil Society." In *Democracy and the Environment: Problems and Prospects,* ed. W. M. Lafferty and J. Meadowcroft. London: Edward Elgar.

Jarvie, I. C. 1972. "The Logic of the Situation." In *Concepts and Society,* 3–36. London: Routledge, Kegan Paul.

Jasanoff, S. 1990. *The Fifth Branch: Science Advisors as Policymakers.* Harvard University Press.

———. 1995. *Science at the Bar: Law, Science, and Technology in America.* Cambridge: Harvard University Press.

Jefferson, Thomas. 1984. *Writings,* ed. Merrill D. Peterson. New York: Library of America.

Jennings, Bruce. 1987. "Policy Analysis: Science, Advocacy, or Counsel?" In *Research in Public Policy Analysis and Management,* ed. S. Nagel. Vol. 4. Greenwich, Conn.: Jai Press.

Jiggins, J. 1989. "An Examination of the Impact of Colonialism in Establishing Negative Values and Attitudes towards Indigenous Agricultural Knowledge." In *Indigenous Knowledge Systems: Implications for Agriculture and International Development,* ed. D. M. Warren et al. Iowa State University, Studies in Technology and Social Change, no. 11. Ames, Iowa: Iowa State University.

Joss, S. 1995. "Evaluating Consensus Conferences: Necessity or Luxury?" In *Public Participation in Science: The Role of Consensus Conferences in Europe,* ed. S. Joss and J. Durant. London: Science Museum, 89–108.

Joss, S. 1998. "Participation in Parliamentary Technology Assessment: From Theory to Practice." In *Parliaments and Technology: The Development of Technology Assessment in Europe,* ed. N. J. Vig and H. Paschen. Albany: State University of New York Press.

Joss, S., and J. Durant, eds. 1995. *Public Participation in Science: The Role of Consensus Conferences in Europe.* London: Science Museum.

Juma, C. 1989. *The Gene Hunters.* London: Zed Press.

Kanigel, R. 1988. "Angry at Our Goods." *Columbia,* October, 23–35.

References

Kann, M. E. 1986. "Environmental Democracy in the United States." In *Controversies in Environmental Policy*, ed. S. Kamieniecki, R. O'Brien, and M. Clarke, 252–74. Albany: SUNY Press.

Kaplan, A. 1998. *The Conduct of Inquiry: Methodology for Behavioral Science*. New Brunswick, N.J.: Transaction.

Kaplan, R. 2000. *The Coming Anarchy*. New York: Random House.

Kaplan, T. 1993. "Reading Policy Narratives: Beginnings, Middles, and Ends." In *The Argumentative Turn in Policy Analysis and Planning*, ed. F. Fischer and J. Forester, 167–85. Durham, N.C.: Duke University Press.

Kapur, A. 1998. "Poor but Prosperous." *Atlantic Monthly*, September, 40–45.

Kassam, Y., and K. Mustafa, eds. 1982. *Participatory Research: An Alternative Methodology in Social Science*. New Delhi: Society for Participatory Research in Asia.

Kassam, Y., and K. Mustafa, eds. 1982. *Participatory Research: An Emerging Alternative in Social Science Research*. Nairobi: African Adult Education.

Kasperson, R., and P. Stallen. 1991. *Communicating Risks to the Public*. Dordrecht, Netherlands: Kluwer.

Kathlene, L., and J. Martin. 1991. "Enhancing Citizen Participation: Panel Designs, Perspectives, and Policy Formation." *Journal of Policy Analysis and Management* 10: 46–63.

Kellert, S. H. 1993. *In the Wake of Chaos: Unpredictable Order in Dynamic Systems*. Chicago: University of Chicago Press.

Kennedy, T. W. 1982. "Beyond Advocacy: A Facilitative Approach to Public Participation." *Journal of the University Film and Video Association* 34, no. 3 (summer): 33–46.

Keren, M. 1995. *Professionals against Populism: The Press, Government, and Democracy*. New York: State University of New York Press.

King, A. 1975. "Overload: The Problems of Governing in the 1970s." *Political Studies* 23: 284–96.

King, G., R. O. Keohane, and S. Verba. 1994. *Designing Social Inquiry: Scientific Inference in Qualitative Research*. Princeton, N.J.: Princeton University Press.

Kingdon, J. 1995. *Agendas, Alternatives, and Public Policy*. New York: HarperCollins.

Kluver, L. 1995. "Consensus Conferences at the Danish Board of Technology." In *Public Participation in Science*, ed. S. Joss and J. Durant, 41–49. London: Science Museum.

Knorr-Cetina, K., and M. Mulkay, eds. 1983. *Science Observed: Perspectives on the Social Study of Science*. London: Sage.

Kotz, N. 1981. "Citizens as Experts." *Working Papers* (March–April): 42–48.

Kraushaar, R. 1985. "Outside the Whale: Progressive Planning and the Dilemmas of Radical Reform." *Journal of the American Planning Association* 54, no. 1 (winter 1988): 91–100.

Krauss, C. 1989. "Community Struggles and the Shaping of Democratic Consciousness." *Sociological Forum* 4, no. 2: 227–39.

Krieger, M. 1981. *Advice and Planning*. Philadelphia, Pa.: Temple University Press.

Krimsky, S. 1984. "Epistemic Considerations on the Values of Folk-Wisdom in Science and Technology." *Policy Studies Review* 3, no. 2: 246–64.

Kritzer, H. M. 1996. "The Data Puzzle: The Nature of Interpretation in Qualitative Research." *American Journal of Political Science* 40, no. 1 (February): 1–32.

Kuhn, T. 1970. *The Structure of Scientific Revolutions*. Chicago: University of Chicago Press.

Kurien, J. 1988. "Knowledge Systems and Fishery Resource Decline: A Historical Perspective." In *Ocean Sciences: Their History and Relation to Man, Proceedings of the Fourth International Congress on the History of Oceanography*, ed. W. Lenz and M. Deacon, 476–80. Hamburg.

Laird, F. N. 1990. "Technocracy Revisited: Knowledge, Power, and the Crisis in Energy Decision Making." *Industrial Crisis Quarterly* 4, no. 1: 49–61.

———. 1993. "Participatory Analysis, Democracy, and Technological Decision Making." *Science, Technology, and Human Values* 18, no. 3: 341–61.

Landy, M. K. 1995. "The New Politics of Environmental Policy." In *The New Politics of Public Policy*, ed. M. K. Landy and M. A. Levin, 207–27. Baltimore: Johns Hopkins University Press.

Landy, M. K., M. J. Roberts, S. R. Thomas. 1994. *The Environmental Protection Agency: Asking the Wrong Questions*. New York: Oxford University Press.

Larson, M. S. 1977. *The Rise of Professionalism*. Berkeley: University of California Press.

Lasswell, H. 1951. "The Policy Orientation." In *The Policy Sciences*, ed. H. Lasswell and D. Lerner. Stanford, Calif.: Stanford University Press.

Latour, B. 1987. *Science in Action*. Cambridge: Harvard University Press.

References

Latour, B., and S. Woolgar. 1979. *Laboratory Life*. Newbury Park, Calif.: Sage.

Laudan, L. 1977. *Progress and Its Problems*. Berkeley: University of California Press.

———. 1984. *Science and Value*. Berkeley: University of California Press.

Lee, D. R. 1990. *Trashing the Planet: How Science Can Help Us Deal with Acid Rain, Depletion of the Ozone, and Nuclear Waste (Among Other Things)*. New York: Harper and Row.

Lee, K. N. 1993. *Compass and Gyroscope: Integrating Science and Politics for the Environment*. Washington, D.C.: Island Press.

Leiss, W. 1990. *Under Technology's Thumb*. Montreal: Queen's University Press.

Lemert, C. 1995. *Sociology after the Crisis*. Boulder, Colo.: Westview Press.

Levine, A. 1982. *Love Canal: Science, Politics, and People*. Boston: Lexington.

Levins, R. 1990. "Toward the Renewal of Science." *Rethinking Marxism* 3: 117.

Lewontin, R. 1997. "Billions and Billions of Demons." *New York Review of Books* (9 January): 29–32.

Liebenburg, L. 1993. "Give Trackers Jobs: Conservation Can Ensure the Survival of Traditional Skill." *New Ground* (spring): 24–26.

———. 1995. *The Art of Tracking: The Origins of Science*. Johannesburg: David Philip Publisher.

Lieberman, J. 1972. *Tyranny of Expertise: How Experts Are Closing the Open Society*. New York: Walker Publishing.

Lillienfeld, A. 1980. *Foundations of Epidemiology*. New York: Oxford.

Lin, A. C. 1999. "Bridging Positivist and Interpretivist Approaches to Qualitative Methods." *Policy Studies Journal* 26, no. 1 (spring): 162–80.

Lin, N. 1976. *Foundations of Social Research*. New York: McGraw-Hill.

Lincoln, Y., and E. Guba. 1985. *Naturalistic Inquiry*. Newbury Park, Calif.: Sage Publications.

Lindblom, C. E. 1990. *Inquiry and Change*. New Haven, Conn.: Yale University Press.

Lindblom, C., and D. Cohen. 1979. *Usable Knowledge: Social Science and Social Problem Solving*. New Haven, Conn.: Yale University Press.

Lindeman, M. 1997. "Building Diversified Deliberative Institutions: Lessons from Recent Research." Paper prepared for presentation at the annual meeting of the American Political Science Association, Washington, D.C., 28 August.

London, S. 1995. "Teledemocracy vs. Deliberative Democracy: A Comparative Look at Two Models of Public Talk." *Journal of Interpersonal Computing and Technology* 3, no. 2 (April): 33–55.

Lopes, L. L. 1987. "The Rhetoric of Irrationality." Paper presented at the Colloquium on Mass Communications, University of Wisconsin, November.

Lowi, T. 1979. *The End of Liberalism*. New York: Norton.

Luke, T. W. 1990. *Screens of Power: Ideology, Domination, and Resistance in Informational Society*. Urbana: University of Illinois Press.

Lynch, M. 1993. *Scientific Practice and Ordinary Action: Ethnomethodology and Social Studies of Science*. Cambridge: Cambridge University Press.

Lynn, L. 1987. *Managing Public Policy*. Boston: Little, Brown.

Lyon, D. 1988. *The Information Society: Issues and Illusions*. Cambridge: Polity Press.

Lyotard, J. 1986. *The Postmodern Condition: A Report on Knowledge*. Minneapolis: University of Minnesota Press.

Maguire, P. 1987. *Doing Participatory Research: A Feminist Approach*. Amherst: Center for International Education, School of Education, University of Massachusetts.

Majone, G. 1989. *Evidence, Argument, and Persuasion in the Policy Process*. New Haven, Conn.: Yale University Press.

Mankiw, N. G. 1977. *Principles of Economics*. New York: Harcourt, Brace, and Jovanovich.

Margolis, H. 1996. *Dealing with Risk: Why the Public and Experts Disagree on Environmental Issues*. Chicago: University of Chicago Press, 1996.

Marrow, A. J. 1969. *The Practical Theorist*. New York: Basic Books.

Martin, B., ed. 1996. *Confronting the Experts*. Albany: State University of New York Press.

Mascarenhas, O., and P. G. Veit. 1994. *Indigenous Knowledge in Resource Management: Irrigation in Msanzi, Tanzania*. Washington, D.C., and Nairobi: World Resources Institute and Acts Press.

Mason, R., and I. Mitroff. 1981. *Challenging Strategic Planning Assumptions*. New York: Wiley.

Mayer, I. *Debating Technologies*. 1997. Tilburg, Netherlands: Tilburg Press.

Mazmanian, D., and D. Morell. 1993. "The 'NIMBY' Syndrome: Facility Siting and the Failure of Democratic Discourse." In *Environmental Policy for the 1990s*, ed. M. Kraft and N. Vig. Washington, D.C.: Congressional Quarterly Press.

References

McAvoy, G. 1999. *Controlling Technocracy: Citizen Rationality and the NIMBY Syndrome*. Washington, D.C.: Georgetown University Press.

McCarthy, T. 1978. *The Critical Theory of Jürgen Habermas*. Cambridge: MIT Press.

McCloskey, D. N. 1985. *The Rhetoric of Economics*. Madison: University of Wisconsin Press.

McCorkle, C. M. 1989. *The Social Science in International Agriculture: Lessons from the CRSPs*. Boulder, Colo.: Lynne Rienner.

McNamara, R. S. 1995. *In Retrospect: The Tragedy and Lessons of Vietnam*. New York: New Times Books.

McNeil, M., ed. 1987. *Gender and Expertise*. London: Free Association Books.

Melnick, S. 1983. *Regulation and the Courts: The Case of the Clear Air Act*. Washington, D.C.: Brookings Institution.

Melucci, A. 1994. "A Strange Kind of Newness: What's 'New' in New Social Movements?" In *New Social Ideology to Identity*, ed. E. Larana et al. Philadelphia, Pa.: Temple University Press.

Menand, L. 1995. "The Trashing of Professionalism." *Academe* 81, no. 3 (May–June): 16–19.

Menkel-Meadow, C. 1995. "The Many Ways of Mediation: The Transformation of Traditions, Ideologies, Paradigms, and Practices." *Negotiation Journal* (July): 217–42.

Merrifeld, J. 1989. *Putting the Scientists in Their Place: Participatory Research in Environmental and Occupational Health*. New Market, Tenn.: Highlander Center.

Michels, R. 1915. *Political Parties*. New York: Dover.

Miller, D. C. 1991. *Handbook of Research Design and Social Measurement*. Newbury Park, Calif.: Sage Publications.

Moore, M. 1983. Working paper, "A Conception of Public Management." Cambridge, Mass.: Kennedy School of Government.

Moore, M. 1995. *Creating Public Value: Strategic Management in Government*. Cambridge: Harvard University Press.

Moore, R., and L. Head. 1993. "Acknowledging the Past, Confronting the Present: Environmental Justice in the 1990s." In *Toxic Struggles: The Theory and Practice of Environmental Justice*, ed. R. Hofrichter, 118–27. Philadelphia, Pa.: New Society.

Moore, S. F. 1985. *Power and Property in Inca, Peru*. Westport, Conn.: Greenwood.

Morrison, R. 1995. *Ecological Democracy*. Boston: South End Press.

Nagel, J. H. 1987. *Participation*. Englewood Cliffs, N.J.: Prentice-Hall.

National Research Council. 1989. *Improving Risk Communication*. Washington, D.C.: National Academy Press.

———. 1996. *Understanding Risk: Informing Decisions in a Democratic Society*. Washington, D.C.: National Academy Press.

Natter, W., T. Schatzku, and J. P. Jones III, eds. 1995. *Objectivity and Its Other*. New York: Guilford.

Nelson, J. S., A. Megill, and D. N. McCloskey. 1986. "Rhetoric of Inquiry." In *The Rhetoric of the Human Sciences*, ed. J. S. Nelson, A. Megill, and D. N. McCloskey, 3–18. Madison: University of Wisconsin Press.

Noble, C. 1987. "Economic Theory in Practice: White House Oversight of OSHA Health Standards." In *Confronting Values in Policy Analysis*, ed. F. Fischer and J. Forester, 266–84. Newbury Park, Calif.: Sage Publications.

Norgaard, R. B. 1984. "Traditional Agricultural Knowledge: Past Performance, Future Prospects, and Institutional Implications." *American Journal of Agricultural Economics* 66: 874–78.

———. 1987. "The Epistemological Basis of Agroecology." In *Agroecology: The Scientific Basis of Alternative Agriculture*, ed. M. Altieri, 21–27. Boulder, Colo.: Westview Press.

Norton, A., and T. Stephens. 1995. *Participation in Poverty Assessment*. Environmental Department Papers, Participation Series. Social Policy and Resettlement Division, World Bank, Washington, June.

Novotny, P. 1994. "Popular Epidemiology and the Struggle for Community Health: Alternative Perspectives from the Environmental Justice Movement." *Capitalism, Nature, and Society* 5, no. 2: 29–42.

———. 1998. "Popular Epidemiology and the Struggle for Community Health in the Environmental Justice Movement." In *The Struggle for Ecological Democracy: Environmental Justice Movements in the United States*, ed. D. Faber, 137–58. New York: Guilford Press.

O'Connor, J. 1993. "The Promise of Environmental Democracy." In *Toxic Struggles: The Theory and Practice of Environmental Justice*, ed. R. Hofrichter, 47–57. Philadelphia, Pa.: New Society.

Offe, C. 1985. "New Social Movements: Challenging the Boundaries of Institutional Politics." *Social Research* 52: 817–68.

Ophuls, W. 1977. *Ecology and the Politics of Scarcity*. San Francisco: Freeman.

Paehlke, R. 1990. "Democracy and Environmentalism: Opening the Door to the Administrative State." In *Managing Leviathan: Environmental Politics and the Administrative State*, ed. R. Paehlke and D. Torgerson, 35–55. Peterborough, N.H.: Broadview Press.

———. 1995. "Environmental Values for a Sustainable Society: The Democratic Challenge." In *Greening Environmental Policy: The Politics of a Sustainable Future*, ed. F. Fischer and M. Black. New York: St. Martin's.

Paehlke, R., and D. Torgerson. 1992. "Toxic Waste as Public Business." *Canadian Public Administration* 35 (fall): 339–62.

Paigen, B. 1982. "Controversy at Love Canal." *Hastings Center Reports* 12, no. 3: 29–37.

Paller, B. 1989. "Extending Evolutionary Epistemology to 'Justifying' Scientific Beliefs." In *Issues in Evolutionary Epistemology*, ed. K. Halweg and C. A. Hooker, 231–57. Albany: SUNY Press.

Paris, D. C., and J. F. Reynolds. 1983. *The Logic of Policy Inquiry*. New York: Longman.

Park, P. 1993. "What Is Participatory Research? A Theoretical and Methodological Perspective." In *Voices of Change: Participatory Research in the United States and Canada*, ed. P. Park et al., 20–46. Toronto: OISE Press.

Parsons, W. 1997. *Keynes and the Quest for a Moral Science: A Study of Economics and Alchemy*. Cheltenham, U.K.: Edgar Elgar.

Participatory Research Network. 1982. *Participatory Research: An Introduction*. New Delhi: Society for Participatory Research in Asia.

Patel, S. 1988. "Enumeration as a Tool for Mass Mobilization: Dharavi Census." *Convergence* 21: 120–35.

Patel, S. J. 1996. "Can the Intellectual Property Rights System Serve the Interests of Indigenous Knowledge?" In *Valuing Knowledge: Indigenous People and Intellectual Property Rights*, ed. S. B. Brush and D. Stabinsky, 305–22. Washington, D.C.: Island Press.

Pateman, C. 1970. *Participation and Democratic Theory*. Cambridge: Cambridge University Press.

———. 1979. *The Problem of Political Obligation*. New York: John Wiley.

Peattie, L. R. 1966. "Reflections on Advocacy Planning." *AIP Journal* (March): 80–88.

Phillips, K. 1989. *The Politics of Rich and Poor*. New York: HarperCollins.

Pierce, J. C., et al. 1992. *Citizens, Political Communication, and Interest Groups: Environmental Organizations in Canada and the United States*. Westport, Conn.: Praeger Publishers.

Piller, C. 1991. *The Fail-Safe Society: Community Defiance and the End of American Technological Optimism*. New York: Basic Books.

Platt, J. 1969. "What We Must Do." *Science* 28 (November): 1117.

Plough, A., and S. Krimsky. 1987. "The Emergence of Risk Communication Studies: Social and Political Context." *Science, Technology, and Human Values* 12, nos. 3–4: 4–10.

Polyani, M. 1983. *Tacit Dimension*. Gloucester, Mass.: P. Smith.

Popper, K. 1959. *The Logic of Scientific Discovery*. London: Heineman Publishers.

Portney, K. E. 1991. *Siting Hazardous Waste Treatment Facilities: The Nimby Syndrome*. New York: Auburn House.

Posey, D. 1990. "Intellectual Property Rights and Just Compensation for Indigenous Knowledge." *Anthropology Today* 6, no. 4: 13–16.

Poster, M. 1990. *The Mode of Information: Poststructuralism and Social Context*. Chicago: University of Chicago Press.

Proctor, R. N. 1991. *Value-Free Science? Purity or Power in Modern Knowledge*. Cambridge: Harvard University Press.

Putt, A. D., and J. F. Springer. 1989. *Policy Research: Concepts, Methods, and Applications*. New York: Prentice-Hall.

Quammen, D. 1997. "Hot and Bothered: An Expert on Climate Change Looks Ahead to the 'Greenhouse Century.'" *New York Times Book Review*, 19 January, 11.

Quantz, R. A. 1992. "On Critical Ethnography (with Some Postmodern Considerations)." In *The Handbook of Qualitative Research in Education*, ed. M. D. LeCompte, W. L. Millroy, and J. Preissle, 447–505. New York: Academic Press.

Rabe, B. G. 1991. "Beyond the Nimby Syndrome in Hazardous Waste Facility Siting: The Albertan Breakthrough and the Prospects for Cooperation in Canada and the United States." *Governance* 4 (April): 184–206.

———. 1992. "When Siting Works, Canada-Style." *Journal of Health Politics, Policy, and Law* 17: 119–42.

———. 1994. *Beyond Nimby: Hazardous Waste Siting in Canada and the United States*. Washington, D.C.: Brookings Institution.

Rahman, M. A. 1991. "Glimpses of the 'Other Africa.'" In *Action and Knowledge: Breaking the Monopoly with Participatory Action-Research*, ed. O. Fals-Borda, 84–108. New York: Apex Press.

Rahman, M. D. A. 1993. *People's Self-Development: Perspectives on Participatory Action Research*. London: Zed Books.

Raloff, J. 1984. "Woburn Survey Becomes a Model for Low-Cost Epidemiology." *Science News*, 18 February.

Rappaport, J., et al. 1985. "Collaborative Research with a Mutual Help Organization." *Social Policy* 15 (winter): 12–17.

Reason, P. 1994. "Three Approaches to Participatory Inquiry." In *Handbook of Qualitative Research*, ed. N. K. Denzin and Y. S. Lincoln, 324–39. London: Sage.

Reason, P., and J. Rowan, eds. 1981. *Human Inquiry: A Sourcebook of New Paradigm Research*. New York: John Wiley.

References

Reich, R. B. 1990. *Public Management in a Democratic Society*. Englewood Cliffs, N.J.: Prentice-Hall.

——, ed. 1988. *The Power of Public Ideas*. Cambridge: Harvard University Press.

Ricci, D. 1984. *The Tragedy of Political Science: Politics, Scholarship, Democracy*. New Haven, Conn.: Yale University Press.

Richards, P. 1979. "Community Environment Knowledge in African Rural Development." *IDS Bulletin* 10, no. 2: 28–36.

Richardson, M., J. Sherman, and M. Gismondi. 1993. *Winning Back the Words: Confronting Experts in an Environmental Public Hearing*. Toronto: Garamond Press.

Ricoeur, P. 1971. "The Model of the Text: Meaningful Action Considered as Text." *Social Research* 38: 529–62.

Rittel, H. W. J., and M. Webber. 1973. "Dilemmas in a General Theory of Planning." *Policy Sciences* 4 (June): 1.

Roberts, A. 1995. " 'Civic Discovery' as a Rhetorical Strategy." *Journal of Policy Analysis and Management*, no. 14 (spring): 291–307.

Roe, E. 1994. *Narrative Policy Analysis*. Durham, N.C.: Duke University Press.

Rorty, R. 1979. *Philosophy and the Mirror of Nature*. Princeton, N.J.: Princeton University Press.

——. 1987. "Science as Solidarity." In *The Rhetoric of the Human Sciences: Language and Argument in Scholarship and Public Affairs*, ed. J. S. Nelson, A. Megill, and D. N. McClosky, 38–52. Madison: University of Wisconsin Press.

Rosenau, P. M. 1992. *Post-modernism and the Social Sciences: Insights, Inroads, and Intrusions*. Princeton, N.J.: Princeton University Press.

Ross, A. 1992. *Strange Weather*. New York: Verso Press.

——. ed. 1996. "Science Wars: Introduction." *Social Text*, nos. 46–47: 1–13.

Rossini, F. A., and A. L. Porter. 1985. "Public Participation and Professionalism in Impact Assessment." In *Citizen Participation in Science Policy*, ed. J. C. Peterson, 62–74. Chicago: University of Chicago Press.

Roszak, T. 1972. *Where the Wasteland Ends: Politics and Transcendence in Postindustrial Society*. Garden City, N.J.: Doubleday.

——. 1994. *The Cult of Information: A Neo-Luddite Treatise on High Tech, Artificial Intelligence, and the True Art of Thinking*. Berkeley: University of California Press.

——. 1995. *The Making of the Counter Culture: Reflections on the Technocratic Society and Its Useful Opposition*. Berkeley: University of California Press.

Rouse, J. 1987. *Knowledge and Power: Toward a Political Philosophy of Science*. Ithaca, N.Y.: Cornell University Press.

Rowell, A. 1996. *Green Backlash: Global Subversion of the Environmental Movement*. London: Routledge.

Rubin, C. T. 1994. *The Green Crusade: Rethinking the Roots of Environmentalism*. New York: Macmillan.

Ruckelshaus, W. D. 1991. "Science, Risk, and Public Policy." In *Taking Sides: Clashing Views on Controversial Environmental Issues*, ed. T. D. Goldfarb. Guilford, Conn.: Duskin.

Sabatier, P., and H. Jenkins-Smith, eds. 1993. *Policy Change and Learning: An Advocacy Coalition Approach*. Boulder, Colo.: Westview.

Saint-Simon, H. 1964. *Social Organization, the Science of Man, and Other Writings*. New York: Harper Torch.

Sale, K. 1995. "Lessons from the Luddites: Setting Limits on Technology." *Nation*, 5 June, 785–88.

Sassower, R. 1997. *Technoscientific Angst*. Minneapolis: University of Minnesota Press.

Savage, G. 1996. *The Social Construction of Expertise: The English Civil Service and Its Influence, 1919–1939*. Pittsburgh, Pa.: University of Pittsburgh.

Schlosberg, D. 1999. *Environmental Justice and the New Pluralism: The Challenge of Difference for Environmentalism*. Oxford: Oxford University Press.

Schmandt, J. 1984. "A New Model of Scientific Regulation." *Science, Technology, and Human Values* 9, no. 1: 23–38.

Schmidt, M. R. 1993. "Grout: Alternative Kinds of Knowledge and Why They Are Ignored." *Public Administration Review* 53, no. 6 (November–December): 525–30.

Schneider, A., and H. Ingram. 1993. "Social Constructions of Target Populations." *American Political Science Review* 87, no. 2: 334–47.

———. 1997. *Policy Design for Democracy*. Lawrence: University Press of Kansas.

Schon, D. 1983. *The Reflective Practitioner*. New York: Basic Books.

Schon, D., and M. Rein. 1994. *Frame Reflection*. New York: Basic Books.

Schram, S. F. 1993. "Postmodern Policy Analysis: Discourse and Identity in Welfare Policy." *Policy Sciences* 26, no. 3: 249–70.

Sclove, R. E. 1995. *Democracy and Technology*. New York: Guilford Press.

Sclove, R., et al. 1998. *Community-Based Research in the United States*. Amherst, Mass.: Loka Institute.

References

Scott, J. C. 1998. *Seeing like a State*. New Haven, Conn.: Yale University Press.

Scott, W. G., and D. K. Hart. 1973. "Administrative Crisis: The Neglect of Metaphysical Speculation." *Public Administration Review* 33 (September–October): 415–22.

Scriven, M. 1987. "Probative Logic." In *Argumentation: Across the Lines of Discipline*, ed. Frans H. van Eemeren et al., 7–32. Dordrecht, Netherlands: Foris Publications.

Selman, P. H. 1997. *Local Sustainability: Managing and Planning Ecologically Sound Places*. New York: St. Martin's Press.

Shah, P. 1995. "Farmers as Analysts, Facilitators, and Decision-Makers." In *Power and Participatory Development*, ed. N. Nelson and S. Wright, 83–94. London: Intermediate Technology Publications.

Sheridan, A. 1980. *Michel Foucault: The Will to Power*. London: Tavistock Publications.

Sherwood, F. 1978. "Action Research and Organizational Learning." *Administration and Society* (December): 21–28.

Sirianni, C., and L. Friedland. 2001. "Civic Innovation in America: Community Empowerment, Public Policy, and the Movement for Civic Renewal." Manuscript. Berkeley: University of California Press, forthcoming.

Skocpol, T. 1985. "Bringing the State Back In: Strategies of Analysis in Current Research." In *Bringing the State Back In*, ed. P. Rueschemeyer and T. Skocpol, 3–43. New York: Cambridge University Press.

Slaton, C. D. 1992. *Televote: Expanding Citizen Participation in the Quantum Age*. New York: Praeger Publishers.

Slovic, P. 1992. "Perception of Risk: Reflections on the Psychometric Paradigm." In *Social Theories of Risk*, ed. S. Krimsky and D. Golding, 117–52. Westport, Conn.: Greenwood.

Slovic, P., et al. 1979. "Rating the Risks." *Environment* 21, 36–39.

Society for Participatory Research in Asia. 1982. *Participatory Research: An Introduction*. Participatory Research Network Series, no. 3. New Delhi: Rajkamal Electric Press.

———. 1987. *Training of Trainers: A Manual for Participatory Training Methodology in Development*. New Delhi: Society for Participatory Research in Asia.

Stake, R. E. 1994. "Case Studies." In *Handbook of Qualitative Research*, ed. N. K. Denzin and Y. S. Lincoln. Newbury Park, Calif.: Sage.

Stanley, M. 1978. *The Technological Consciousness: Survival and Dignity in an Age of Expertise*. Chicago: University of Chicago Press.

Stern, P. C., and H. V. Feinberg. 1996. *Understanding Risk: Informing*

Decisions in a Democratic Society. Washington, D.C.: National Academy Press.

Stewart, T. R., R. L. Dennis, and D. W. Ely. 1984. "Citizen Participation and Judgment in Policy Analysis: A Case Study of Urban Quality." *Policy Sciences* 17: 67–87.

Stockman, N. 1983. *Anti-positivist Theorists of the Sciences: Critical Rationalism and Scientific Realism*. Dordrecht, Netherlands: D. Reidel.

Stone, D. A. 1988. *Policy Paradox and Political Reason*. Glenview, Ill.: Scott, Foresman.

Stull, D., and J. Schensul, eds. 1987. *Collaborative Research and Social Change*. Boulder, Colo.: Westview Press.

Sullivan, W. M. 1983. "Beyond Policy Science." In *Social Science as Moral Inquiry*, ed. N. Haan, P. Rabinov, and W. M. Sullivan, 297–319. New York: Columbia University Press.

Susman, G. I. 1983. "Action Research: A Sociotechnical Systems Perspective." In *Beyond Method*, ed. Gareth Morgan, 95–113. Beverly Hills, Calif.: Sage.

Susskind, L., and C. Ozawa. 1985. "Mediating Public Disputes: Obstacles and Possibilities." *Journal of Social Issues* 41, no. 2: 145–59.

Sylvia, R. D., et al. 1991. *Program Planning and Evaluation for the Public Manager*. Prospect Heights, Ill.: Waveland.

Tandon, R. 1988. "Social Transformation and Participatory Research." *Convergence* 21: 5–18.

Taylor, P., and F. H. Buttel. 1992. "How Do We Know We Have Global Environmental Problems? Science and the Globalization of Environmental Discourse." *Geoforum* 23.

Tesh, S. N. 1999. "Citizen Experts in Environmental Risk." *Policy Sciences* 32, no. 1: 39–58.

Thakur, R. 1995. *The Government and Politics of India*. New York: St. Martin's Press.

Thornton, J. 1991. "Risking Democracy." *Greenpeace*, March–April, 14–17.

Thrupp, L. A. 1984. "Women, Wood, and Work: In Kenya and Beyond." *FAO Journal of Forestry* (December).

———. 1989. "Legitimizing Local Knowledge: From Displacement to Empowerment for Third World People." *Agriculture and Human Values* 6, no. 3 (summer): 13–24.

Toffler, A. 1991. *The Third Wave*. London: Pan.

———. 1993. *Future Shock*. London: Heyday.

Torbert, W. 1976. *Creating a Community of Inquiry: Conflict, Collaboration, Transformation*. New York: John Wiley.

———. 1981. "Empirical, Behavioral, Theoretical, and Attentional Skills

Necessary for Collaborative Inquiry." In *Human Inquiry: A Sourcebook of New Paradigm Research*, ed. P. Reason and J. Rowan. Chichester, U.K.: John Wiley.

———. 1991. *The Power of Balance: Transforming Self, Society, and Scientific Inquiry*. Newbury Park, Calif.: Sage.

Toulmin, S. 1958. *The Uses of Argument*. Cambridge: Cambridge University Press.

———. 1983. "The Construal of Reality: Criticism in Modern and Postmodern Science." In *The Politics of Interpretation*, ed. W. J. T. Mitchell, 99–117. Chicago: University of Chicago Press.

———. 1990. *Cosmopolis: The Hidden Agenda of Modernity*. Chicago: University of Chicago Press.

Touraine, A. 1965. *Sociologie de l' action*. Paris: Editions du Seuil.

———. 1981. *The Voice and the Eye*. Cambridge: Cambridge University Press.

Tribe, L. 1972. "Policy Science: Analysis or Ideology." *Philosophy and Public Affairs* 2: 66–110.

Uchitelle, L. 1999. "A Challenge to Scientific Economics." *New York Times*. January 23, B7.

Uphoff, N. 1992. *Learning from Gal Oya: Possibilities for Participatory Development and Post-Newtonian Social Science*. Ithaca, N.Y.: Cornell University Press.

van den Daele, W. 1995. "Technology Assessment as a Political Experiment." In *Contested Technology: Ethics, Risk, and Public Debate*, ed. Rene von Schomberg. Tilburg: Center for Human and Public Affairs.

van der Ploeg, J. 1989. "Knowledge Systems, Metaphor, and Interface: The Case of Potatoes in the Peruvian Highlands." In *Encounters at the Interface: A Perspective on Social Discontinuities in Rural Development*, ed. N. Long. Wageningse Sociologische Studies, 27. Landbouw Universisteit Wagening.

van der Ploeg, J. D. 1993. "Potatoes and Knowledge." In *An Anthropological Critique of Development: The Growth of Ignorance*, ed. M. Hobart, 209–27. London: Routledge.

Wagner, P. 1995. "Sociology and Contingency: Historicizing Epistemology." *Science Information* 34, no. 2: 179–204.

Wagner, P., C. Weiss, B. Wittrock, and H. Wollmann, eds. 1991. *Social Science and Modern States: National Experiences and Theoretical Crossroads*. Cambridge: Cambridge University Press.

Wallerstein, I., et al. 1996. *Open the Social Sciences: Report of the Gulbenkian Commission on the Restructuring of the Social Sciences*. Stanford, Calif.: Stanford University Press.

Wartenburg, D. 1989. "Quantitative Risk Assessment." *Science for the People* (January–February): 18–23.

Watson-Verran, H., and D. Turnbull. 1994. "Science and Other Indigenous Knowledge Systems." In *Handbook of Science and Technology Studies*, ed. S. Jasanoff et al., 115–39. Newbury Park, Calif.: Sage.

Webler, T. 1999. "The Craft and Theory of Public Participation: A Dialectical Process." *Journal of Risk Research* 2, no. 1: 55–71.

Weiss, C. 1990. "Policy Research: Data, Ideas, or Arguments? In *Social Sciences and Modern States*, ed. P. Wagner et al. Cambridge: Cambridge University Press.

Welsh, I. 1993. "The NIMBY Syndrome and Its Significance in the History of the Nuclear Debate in Britain." *British Journal of the History of Science* 26, no. 1: 15–32.

———. 1995. *Nuclear Power: Generating Dissent*. London: Routledge.

Whitaker, D., L. Archer, and S. Greve. 1990. *Research, Practice, and Service Delivery: The Contribution of Research by Practitioners*. London: Central Council for Education and Training in Social Work.

Whyte, W. F. 1989. "Advancing Scientific Knowledge through Participatory Action Research." *Sociological Forum* 4: 367–86.

Wildavsky, A. 1988. *The Search for Safety*. New Brunswick, N.J.: Rutgers University Press.

———. 1997. *But Is It True? A Citizen Guide to Environmental Health and Safety Issues*. Cambridge: Harvard University Press.

Willard, C. A. 1996. *Liberalism and the Problem of Knowledge: A New Rhetoric for Modern Democracy*. Chicago: University of Chicago Press.

Williams, B. 1985. *Ethics and the Limits of Philosophy*. Cambridge: Harvard University Press.

Wineman, S. 1984. *The Politics of Human Services: A Radical Alternative to the Welfare State*. Boston: South End Press.

Winner, L. 1977. *Autonomous Technology: Technics-Out-of-Control as a Theme in Political Thought*. Cambridge: MIT Press.

———. 1986. *The Whale and the Reactor: A Search for Limits in an Age of High Technology*. Chicago: Chicago University Press.

Withhorn, A. 1984. *Serving the People: Social Services and Social Change*. New York: Columbia University Press.

Woolgar, S. 1988. *Science—The Very Idea*. London: Tavistock.

Woozley, A. D. 1949. *Theory of Knowledge*. London: Hutchinson.

World Bank. 1994. *The World Bank and Participation, Operations Policy Department*. World Bank, September.

———. 1995. *World Bank Participation Sourcebook*. Environmental Department Papers. World Bank, June.

References

Wynne, B. 1987. *Risk Management and Hazardous Wastes: Implementation and the Dialectics of Credibility*. Berlin: Springer.

———. 1992. "Misunderstood Misunderstanding: Social Identities and Public Uptake of Science." *Public Understanding of Science* 1, no. 3: 231–304.

———. 1996. "May the Sheep Safely Graze? A Reflexive View of the Expert-Lay Knowledge Divide." In *Risk, Environment, and Modernity: Toward a New Ecology*, ed. S. Lash, B. Szerzynski, and B. Wynne, 45–80. Newbury Park, Calif.: Sage.

Yankelovich, D. 1999. *The Magic of Dialogue: Transforming Conflict into Cooperation*. New York: Simon and Schuster.

Yanow, D. 1996. *How Does a Policy Mean? Interpreting Policy and Organizational Actions*. Washington, D.C.: Georgetown University Press.

———. 1998. *Conducting Interpretive Policy Analysis*. Newbury Park, Calif.: Sage.

Yearly, S. 1992. *The Green Case: A Sociology of Environmental Issues, Arguments, and Politics*. London: Routledge.

Index

Frank Fischer is Professor of Political Science at Rutgers University on the Newark campus and a member of the Bloustein School of Planning and Public Policy in New Brunswick. He is author of various books, including *Evaluating Public Policy* (1995), *Technocracy and the Politics of Expertise* (1990), *Politics, Values, and Public Policy* (1980). He has also edited a number of titles, including (with Maarten A. Hajer) *Living with Nature: Environmental Politics as Cultural Discourse* (1999) and (with John Forester) *The Argumentative Turn in Policy Analysis and Planning* (1993).

Library of Congress Cataloging-in-Publication Data

Fischer, Frank, 1942–
Citizens, experts, and the environment : the politics of local knowledge / Frank Fischer.
p. cm.
Includes bibliographical references and index.
ISBN 0-8223-2628-0 (cloth : alk. paper) —
ISBN 0-8223-2622-1 (pbk. : alk. paper)
1. Environmental policy — Citizen participation. I. Title.
GE170.F52 2000
363.7'0525 — dc21   00-029398